POLYPHASE
COMMUTATOR MACHINES

POLYPHASE
COMMUTATOR MACHINES

BY

B. ADKINS
M.A., M.I.E.E., A.M.I.C.E.

AND

W. J. GIBBS
D.SC., M.I.E.E.

CAMBRIDGE
AT THE UNIVERSITY PRESS
1951

CAMBRIDGE UNIVERSITY PRESS
Cambridge, New York, Melbourne, Madrid, Cape Town,
Singapore, São Paulo, Delhi, Tokyo, Mexico City

Cambridge University Press
The Edinburgh Building, Cambridge CB2 8RU, UK

Published in the United States of America by Cambridge University Press, New York

www.cambridge.org
Information on this title: www.cambridge.org/9780521233514

First published 1951
First paperback edition 2011

A catalogue record for this publication is available from the British Library

ISBN 978-0-521-23351-4 Paperback

PREFACE

There has been for a long time a need for a book on polyphase commutator machines, written by designers and intended for students and for engineers interested in their application. These machines are widely used in spite of their extra complication and cost, because they do overcome the limitations of the ordinary induction motor. Commutator machines are mainly used where the induction motor by itself is unsuitable, because of a need for variable speed or high power factor. Large numbers of all sizes have been in successful operation for many years, and there is still an important and expanding field of application.

The book is laid out as far as possible from a practical point of view. It describes, for each type of machine, the most important methods of control and the characteristics obtained. The theory is given in separate sections which can be omitted by those readers who are interested in what a machine will do rather than how it works.

It is essential for the reader who wishes to follow the theoretical treatment to have a good prior knowledge of induction motor theory. The treatment of this theory given in the first three sections of Chapter 1 is included in order to introduce the fourth section, which deals with the behaviour of an induction motor when a voltage is 'injected' into its secondary, the later chapters giving the theory of commutator machines. We believe that the subject of polyphase commutator machines is by no means as difficult as it is often thought to be, because it follows naturally from induction motor theory. This becomes more evident when the subject is treated as a whole instead of piecemeal. The unified analytical theory given in this book is more concise than most earlier treatments.

The polyphase commutator machines always arouse interest among students who study and test them, and among engineers who use them in industry; the demand for information on the subject is widespread, and this book is intended to satisfy that demand. It embodies the results of many years spent in the design of such machines and the investigation of their problems.

We are indebted to the Directors of the British Thomson-Houston Company Ltd. for permission to publish this work. We also acknowledge with thanks the many valuable suggestions made by Mr R. A. Hayes, M.A., former Fellow of Trinity Hall, Cambridge, and the helpful criticisms and assistance in reading the proofs from our colleague Mr E. C. Barwick, B.Sc.

B. ADKINS

W. J. GIBBS

RUGBY

April 1948

CONTENTS

PHOTOGRAPHS OF MACHINES
(at the end of the book)

INTRODUCTION

Many types of electrical machine are used in connexion with polyphase alternating current supply systems, both as generators for supplying the electrical power, and as motors for converting it into mechanical power. The commutator machines occupy a special place, since, in return for some additional complication, they can satisfy requirements which cannot be met by the simpler and commoner machines.

Polyphase commutator machines comprise those types of electrical rotating machine which have a commutator, and in which the commutator brush gear carries polyphase alternating currents. There are many types of such machines, but they have certain common features which give an underlying unity to the subject taken as a whole. The definition given above excludes single-phase commutator machines, which have in general different features, and form a separate subject. It also excludes machines in which the commutator carries direct current, such as the self-excited synchronous generator, the synchronous induction motor, and the rotary converter of a Kramer equipment. The important common feature of polyphase machines is that the general operation of all types can be explained in terms of a rotating field similar to that of the induction motor.

The polyphase commutator machine can, in general, be used either:

(a) as an independent motor, generator, or converter;

or (b) as a regulating machine connected in the rotor circuit of an induction motor, for the purpose of regulating the speed or power factor of the induction motor.

With either method of application, the results obtained depend not only on the construction of the machine, but on the connexions between its own windings, with other apparatus, and with the supply.

Each machine has stator and rotor cores forming a magnetic circuit, and two or more polyphase windings fitted in slots around the periphery of stator and rotor. From the point of view of the construction of the machine itself and its windings, the polyphase commutator machines can be classified into three general groups, as explained later. Inside each group, many variations are possible, by such means as providing additional windings or altering the external connexions.

From the theoretical point of view a still greater unity of treatment is possible. The theory of most types of polyphase commutator machine, including all the important ones, can be derived from the theory of a slip-ring induction motor which has a slip-frequency voltage applied to, or 'injected' into, the secondary circuit. In order to develop this method of treatment, it is necessary to consider how the characteristics of an induction motor can be modified by means of an injected voltage, irrespective of how this voltage is produced. This forms a basis for the study of nearly all types of commutator machine, including both motors and regulating machines. The only exceptions are those of minor importance, where the commutator machine is used as an independent source of low-frequency power.

Historical notes

Quite early in the history of electricity supply, it was recognized that the generation of alternating current was both easier and more fundamental than the generation of direct current; but since the application of alternating current to the driving of machinery was not so simple, the first practical installations were operated with direct current. Attempts were made to run the D.C. motor, or modifications of it, from a single-phase supply, in order to obtain the benefits of A.C. generation. When polyphase alternating current was introduced, new types of commutator motor were developed for use on the two- or three-phase supplies.

Just after 1888, and about the time that Tesla developed the induction motor, Görges brought into use the series polyphase motor. A little later, various types of phase advancers and slip regulators were introduced, for use with slip-ring induction motors. In 1901, Roth and others developed the polyphase shunt motor, using a tapped transformer for controlling the speed, and in 1906, Scherbius and Lydall introduced the control system which uses a commutator machine provided with interpoles, and which was afterwards developed further by Hull for double-range operation. Finally, Schrage took out the patent for his variable-speed A.C. motor in 1912.

By then, all the fundamental types of to-day were known, and subsequent progress has depended on details of design and application. The most important recent development has been the introduction of various types of damping winding for assisting commutation in machines not provided with interpoles.

From the application point of view the changes have been considerable. It has been increasingly recognized that there are very few duties where a motor with series characteristics is preferable to one with shunt characteristics; consequently the series motor, although the first type of polyphase commutator motor to be built, is now comparatively little used. In the steel industry, which furnished many of the early applications, the demand for larger units with very wide speed ranges has led to the use of Ward Leonard D.C. drives rather than A.C. commutator machines. On the other hand, there has been an enormous increase in the use of commutator machines of all sizes in other industries, and the range of motors has been extended down to very small sizes, even below 1 H.P., sizes which were previously thought to be uneconomical for this type of machine.

Developments in the theory of polyphase commutator machines have followed similar lines to those relating to polyphase induction motors. In the early days, each type was analysed independently by means of geometrical methods similar to, but more complicated than, those used for the induction motor. More recent study of the subject has shown the advantage of analytical methods; these not only are more convenient for the actual computation work, but they bring out the similarity between different types, and make comparisons between them easier.

Plan of the book

The first five chapters are largely devoted to general questions which underlie the theory of operation of all types. In Chapter 1, the theory of the induction motor is extended to cover the application of an injected voltage to the secondary. This is followed in Chapter 2 by a study of the windings used in commutator machines, and in Chapter 3 the machines are classified into three main groups, which are defined in terms of the windings used. Chapters 4 and 5 deal with commutation and other special problems common to all types.

Some details about the construction and general arrangement of each type of machine are given in Chapter 3, and each of the three groups is discussed fully in Chapters 6, 7 and 8. These chapters contain, for each type, a description of the practical arrangements for controlling the machines, an explanation of the theory necessary for predetermining the performance, and a summary of the characteristics obtained. Chapter 9 deals with the special modifications used in phase advancers and other machines when power-factor correction is required without speed variation. The last chapter is concerned with the many practical applications of these machines.

In applying the theory of the induction motor with an injected voltage to each of the types of commutator machine, suitable modifications have to be made. For the shunt commutator motor, the only important change is in the manner of variation of the secondary reactance. In the treatment of the Schrage motor, the magnetic linkage between the primary and regulating windings must be taken into account. In the series motor the voltage applied to the primary is variable, and the current is common to both stator and rotor. For Scherbius control schemes, additional relationships must be included, depending on the connexions of the field windings of the regulating machine.

At first sight it may appear that the attempt to bring all these machines into a common framework is somewhat artificial, particularly as it has been usual in the past to treat each type of machine independently on its own merits. But experience in designing a wide range of different machines has shown that the attempt to predetermine the performance of any type, can always be based on the theory of the polyphase induction motor, and the methods described have been successfully used for this purpose. This line of approach is also useful for the non-specialist, because the induction motor theory is very widely understood, and the method of treatment makes it easier for him to gain an understanding of the properties of the various types of commutator machine.

Chapter 1

SPEED AND POWER-FACTOR REGULATION OF INDUCTION MOTORS

1. Vector diagrams of the induction motor

The fundamental principle underlying the operation of nearly all polyphase commutator machines is the fact that the speed and power factor of an induction motor can be controlled at will by applying an external voltage of the correct frequency to the secondary circuit. An understanding of the theory of operation of the induction motor is therefore of prime importance, since it serves as a basis for subsequent theory.

The polyphase induction motor has a primary winding into which the power flows from the line. In slip-ring and squirrel-cage induction motors, the primary winding is nearly always situated on the stator, and that arrangement is assumed in the present chapter. It can, however, equally well be on the rotating part; the theory is not affected in any way by interchanging the positions of primary and secondary, because the operation depends only on the relative motion.

The winding is distributed round the stator periphery in such a way that the coils of the various phases follow each other in space in the same way in which the currents and voltages in the phases follow each other in time. The resulting air-gap flux of the machine is taken to be sinusoidally distributed in space, and to rotate round the air-gap at a speed which depends on the frequency of the supply and the number of poles for which the stator is wound. This speed is called the *synchronous speed*, n_0, and for p poles and a supply frequency f cycles per second

$$n_0 = \frac{2f}{p} \text{ rev./sec.} \tag{1}$$

The actual speed of the rotor is denoted by n.

The magnitude of the flux is determined by the fact that the voltage it induces in the primary winding must balance the supply voltage after deducting the drop due to primary resistance and leakage reactance as defined below. Under ordinary load conditions, with constant supply voltage, it is therefore an approximately constant flux, independent of the current flowing, except in so far as the current produces resistance or reactance voltage drops.

The total flux in the machine is considered to consist of three parts:

(1) The *main flux* Φ, linking both stator and rotor windings.

(2) The *primary leakage flux* which links the primary but not the secondary windings. Neglecting saturation effects, this flux is proportional to the primary current, and has the same effect as a reactance X_1—called the *primary leakage reactance*—connected in series with each phase of the primary winding.

(3) The *secondary leakage flux* which links the secondary but not the primary windings. Again neglecting saturation effects, it is proportional to the secondary current and has the same effect as a reactance in series with the secondary winding. This *secondary leakage reactance* is proportional to the variable secondary frequency, which is equal to that of the supply when the rotor is at rest but is less than the supply frequency when the rotor runs below synchronous speed in the direction of the field. The secondary leakage reactance at supply frequency is denoted by X_r.

The stator and rotor resistances are denoted by R_1 and R_r respectively.

Induced voltages

The voltage induced in a phase of the primary winding by the main flux is the *primary induced voltage* of magnitude E. The voltage induced in a secondary phase by the main flux is the *secondary induced voltage*.

The magnitude of the secondary induced voltage depends on the relative speed $(n_0 - n)$ between the flux and the rotor. The value at standstill is denoted by E_r. The ratio of the relative speed to the synchronous speed is called the *slip*, and is denoted by s. Thus

$$s = \frac{n_0 - n}{n_0}. \tag{2}$$

Hence at speed n, the secondary induced voltage is sE_r, the secondary frequency is sf, and the secondary leakage reactance is sX_r.

Vector diagram of secondary voltages

In this section the normal induction motor with short-circuited secondary, operating below synchronous speed, is considered. Under such a condition, the secondary induced voltage sE_r causes the current I_r to flow through the secondary impedance; the current I_r, by interaction with the main flux, produces the torque of the machine.

Fig. 1 shows the vector diagram of voltages in one secondary phase. It consists of a triangle in which sE_r is equal to the vector sum of the resistance drop I_rR_r, and the leakage reactance drop sI_rX_r. In this diagram the vectors are rotating at slip frequency in the direction indicated by the arrow. At small values of slip the leakage reactance sX_r is relatively small, and it causes the secondary current I_r to lag behind the voltage by the small angle ϕ_2; $\cos \phi_2$ is the secondary power factor.

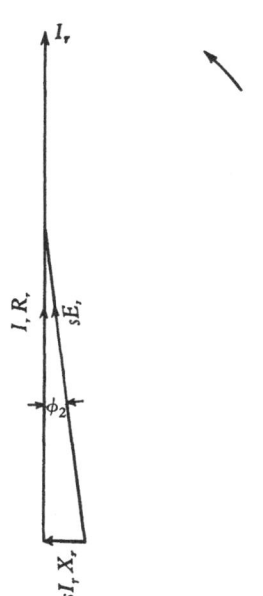

Fig. 1. Vector diagram of secondary voltages

Currents and magnetomotive forces

The current flowing in any winding sets up a magnetomotive force; the reluctance of the path of the main flux is assumed to be concentrated at the air-gap, and hence the value of the M.M.F. set up is also concentrated at the air-gap. The distribution of M.M.F. can thus be represented by a function of position around the air-gap. Using this conception, the currents flowing in the two windings set up two space waves of M.M.F. both of which rotate at synchronous speed. The primary and secondary M.M.F. waves combine to give the magnetizing M.M.F. which produces the flux Φ. A secondary current and its M.M.F. are taken to be positive when the resultant M.M.F. is the difference between primary and secondary M.M.F.'s.

Fig. 2 is a vector diagram of currents representing the M.M.F.'s. In order to represent the M.M.F.'s by currents on the same vector diagram, a primary current I_2 is used instead of the slip-frequency secondary current I_r. I_2, which is called the *referred secondary current*, is the primary current whose M.M.F. is equal to the M.M.F. due to the actual secondary current. The magnetizing current I_0 is then equal to the vector difference between the primary current I_1 and the referred secondary current I_2 as shown in Fig. 2.

The component of the supply voltage which is required to balance the actual primary induced voltage is called the *internal voltage*, and is represented in Fig. 2 by the vector E. If there were no losses due to hysteresis or eddy currents, I_0 would lag behind E by 90°, but in practice the angle is less than 90° by a small angle ψ. In Fig. 2 the two components of I_0 are shown. I_m is the true magnetizing component lagging 90° behind E, and I_c is the core loss component in phase with E.

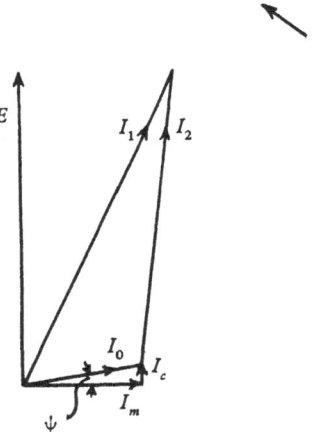

Fig. 2. Vector diagram of primary currents

The angle by which the referred secondary current I_2 lags behind the internal voltage E must be the same as the angle ϕ_2 by which the actual secondary current I_r lags behind the secondary induced voltage sE_r, because the currents set up the same M.M.F. and the primary and secondary induced voltages are due to the same flux. Hence the phase of I_2 is determined by Fig. 1, and the angle between I_2 and E is ϕ_2.

Vector diagram of primary voltages

In Fig. 3, which shows the vector diagram of primary voltages, the supply voltage is equal to the sum of the internal voltage E and the primary impedance drop, that is, the resistance drop I_1R_1 and the leakage reactance drop I_1X_1. As in Fig. 2 the vectors rotate in the direction of the arrow at supply frequency. The primary current lags behind the supply voltage by the angle ϕ and the power factor is $\cos\phi$.

Equivalent unity ratio motor

The magnitudes of the primary induced voltage E and the standstill secondary induced voltage E_r depend on the effective number of turns in a phase of the winding, the effective number of turns being defined as the actual number of turns in series, multiplied by the winding coefficient. The ratio of E to E_r is equal to the ratio of the *primary effective turns* T_1 to *secondary effective turns* T_r, and is called the *transformation ratio* K:

$$K = \frac{T_1}{T_r} = \frac{E}{E_r}. \qquad (3)$$

If the numbers of effective turns in stator and rotor, and the numbers of phases were the same, E_r would be equal to E, and I_2 would be equal to I_r. In practice the numbers of effective turns are different, and even the numbers of phases sometimes differ. In treating the theory, however, it is very convenient to deal with an equivalent motor in which the numbers of effective turns are the same in primary and secondary. The performance of the equivalent unity ratio motor is the same as that of the actual motor.

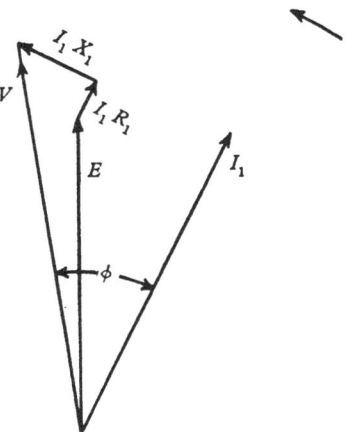

Fig. 3. Vector diagram of primary voltages

The secondary current of the equivalent motor is equal in magnitude (though of different frequency) to the referred secondary current I_2. Since I_2 produces the same M.M.F. as I_r

$$I_2 = \frac{m_r}{mK} I_r, \qquad (4)$$

where m is the number of stator phases, and m_r is the number of rotor phases.

The *referred secondary voltage* is $sE = sKE_r$.

The secondary resistance and reactance of the equivalent motor also differ from those of the actual motor. They are the *referred values*, and since each is a voltage divided by a current, they are given by

$$\textit{Referred secondary resistance } R_2 = \frac{m}{m_r} K^2 R_r. \qquad (5)$$

$$\textit{Referred secondary reactance } sX_2 = \frac{m}{m_r} K^2 s X_r. \qquad (6)$$

It should be clearly understood that the transformation used when referring values of secondary voltage, current, resistance and reactance, from the actual motor to the equivalent motor brings in ratios of effective turns and numbers of phases only.

Vector diagram of referred secondary voltages

The vector diagram of secondary voltages is drawn again in Fig. 4 using referred values. The diagram may be considered either as a supply frequency diagram in which the voltage and current are referred to corresponding values in the primary, or as a slip-frequency diagram representing the secondary currents of the equivalent motor. In either case, if the slip is positive, the vectors rotate counter-clockwise as shown by the arrow. This point is discussed further in Section 4.

In the theory which follows, the secondary values of voltage, current, resistance, and reactance are the referred values and not actual values. When making calculations the first step is to determine the referred values of the secondary quantities, and for the sake of brevity the word 'referred' is omitted in later sections. Similarly, leakage reactance is called simply reactance, unless otherwise stated.

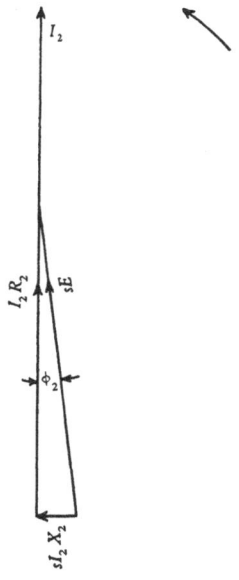

Fig. 4. Vector diagram of referred secondary voltages

Thus

I_2 = secondary current,

sE = secondary induced voltage,

R_2 = secondary resistance,

sX_2 = secondary reactance, all represent referred values.

The current taken by the motor at any given speed is therefore determined by the following considerations. The voltage induced in the primary winding must balance the applied voltage less the primary impedance drop. This determines the flux Φ and the magnetizing current I_0. The voltage induced in the secondary winding is proportional to the flux and to the slip, and in the equivalent unity ratio motor the secondary current I_2 is the current passed through the secondary impedance by this voltage. Since the primary current I_1 is the vector sum of I_0 and I_2, all the currents are thereby determined for any value of slip.

There are certain assumptions underlying both the vector diagrams and the subsequent theory based on them. Chief of these is the assumption that all the voltages and currents are sinusoidal in time, and that the M.M.F.'s and flux are sinusoidally distributed in space and rotate at a uniform speed. In practice this means that the basic theory deals with fundamental components only, and that the effects of harmonics are neglected. Further, to simplify the analysis, it is necessary to assume that all relationships between voltages and currents are linear. In practice, the reactances are not constant because the iron permeability varies, and resistances are not constant because the windings have different temperatures at different loads. The use of the current I_c assumes that all the iron loss originates in the stator iron. Moreover, hysteresis introduces a special effect because the torque it produces reverses when the motor passes through synchronous speed. This small hysteretic torque is neglected throughout the book. All these simplifications are necessary because the problems of an actual machine are much too involved for complete rigour of analysis; what is required is a theory as nearly rigorous as is reasonable. The vector diagrams given in this section form a sound basis for practical calculations of the performance of induction motors. Other factors, such as the effects of harmonics, have to be treated separately.

2. Equivalent circuit and equations of the induction motor

The vector diagrams in Section 1 provide a complete theory of the induction motor, and can be used to determine its characteristics. However, it is not very convenient to work directly from the vector diagrams, and it is usual to derive from it an equivalent circuit which gives the required result more easily. The performance of the motor is then deduced from the equivalent circuit either by an analytical method using equations in which the vectors are represented by complex quantities, or by a graphical method such as the well-known 'circle diagram'.

Equivalent circuit with ideal induction motor

The equivalent circuit of the induction motor is the same as that of the transformer, except for the important difference that, because of the rotation of the rotor, the secondary voltage induced by a given flux is not constant but varies with the slip. The first step in deriving the equivalent circuit is to introduce the notion of an 'ideal' induction motor, which is analogous to the ideal transformer used in transformer theory.

The *ideal induction motor* is one in which the transformation ratio is unity, the number of secondary phases is the same as the number of primary phases, and there is no magnetizing current, core loss, resistance or leakage reactance. As stated in Section 1, the practical motor has the same characteristics as an equivalent unity ratio motor. The equivalent unity ratio motor, for which the vector diagrams in Figs. 2, 3 and 4 apply, is in turn equivalent to an ideal induction motor to which external resistances and reactances have been added as in the circuit diagram Fig. 5. This circuit therefore takes the same current I_1 from the supply as the actual induction motor. Like the vector diagrams, the equivalent circuit applies to one phase.

The primary voltage of the ideal motor is the internal voltage E of the actual motor. The secondary induced voltage is sE, and this causes the secondary current I_2 to flow through the secondary impedance Z_2, comprising the resistance R_2 and the reactance sX_2. The primary current of the ideal motor is also I_2, and to obtain the primary current I_1 of the actual motor, the current I_0 must be added to it. In Fig. 5 a circuit consisting of the magnetizing reactance X_0 and the core loss resistance R_0 in parallel is connected across the ideal motor. I_0 is the vector sum of the true magnetizing current I_m in the reactance X_0, and the core loss current I_c in the resistance R_0; the total primary current I_1 is the vector sum of I_2 and I_0. Finally, to obtain the applied voltage V, the impedance drop due to I_1 in the primary impedance Z_1 which comprises the resistance R_1 and the reactance X_1, must be added to the internal voltage E. It is evident that the vector diagrams of Figs. 2–4 apply to this circuit.

The secondary voltages and currents of the ideal motor are, of course, of slip frequency. However, it is possible to assign to the element designated 'ideal motor', the property of converting actual slip-frequency voltages and currents to the equivalent supply frequency values for which the same vector diagrams apply. These alternative conceptions are discussed further on pp. 22 and 27.

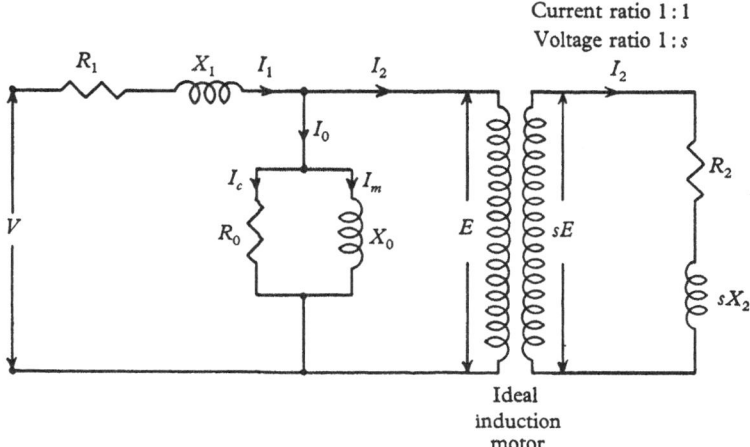

Fig. 5. Equivalent circuit including ideal induction motor

Equations of the induction motor with short-circuited secondary

From the equivalent circuit of Fig. 5, equations relating the quantities can be written down, and from these equations can be calculated not only the primary current I_1 but also the secondary current I_2, the internal voltage E, and the torque T. All the currents and voltages are vectors, having both magnitude and phase, and are represented in the equations by complex quantities.

In the equations:

$$\text{Primary impedance} \qquad Z_1 = R_1 + jX_1, \qquad (7a)$$

$$\text{Secondary impedance} \qquad Z_2 = R_2 + jsX_2, \qquad (7b)$$

$$\text{Magnetizing admittance} \quad Y_0 = G_0 - jB_0, \qquad (7c)$$

where $G_0 = 1/R_0$, $B_0 = 1/X_0$. Then

$$I_0 = EY_0, \qquad (8a)$$

$$I_1 = I_0 + I_2, \qquad (8b)$$

$$V = E + I_1 Z_1, \qquad (8c)$$

$$sE = I_2 Z_2. \qquad (8d)$$

Combining (8a), (8b) and (8c)

$$V = E(1 + Y_0 Z_1) + I_2 Z_1$$
$$= cE + I_2 Z_1, \qquad (9)$$

where $\qquad c = 1 + Y_0 Z_1.$

Combining (8a) and (9)

$$I_0 = \frac{V Y_0}{c} - \frac{I_2 Y_0 Z_1}{c}. \qquad (10)$$

Also, from (8d) and (9)

$$I_2 = \frac{V}{Z_1 + c Z_2 / s}$$
$$= \frac{sV}{s Z_1 + c Z_2}, \qquad (11)$$

and from (8b) and (10)

$$I_1 = \frac{V Y_0}{c} - \frac{I_2 Y_0 Z_1}{c} + I_2$$
$$= \frac{V Y_0}{c} + \frac{I_2}{c}. \qquad (12)$$

The introduction of the quantity c simplifies the calculations considerably, particularly for the more complicated circuits considered later. Strictly, c is a complex quantity, but in practice the imaginary part is so small that there is no appreciable error in treating c as a real quantity. It should be noted that on no-load, when $I_2 = 0$,
$$V = cE,$$
c is thus the ratio of applied voltage to internal voltage at no-load and is for all practical purposes a real quantity slightly greater than unity.

Since the rotor winding is short-circuited, the input W_1 to the motor primary is equal to the mechanical output plus the losses. If the stator copper and core losses are deducted from the input, the remainder is the power transmitted to the rotor across the gap by the rotating flux, and is known as the *synchronous power* or *synchronous watts* W_T of the motor. The input power W_1 and the synchronous watts are given by

$$W_1 = m V I_1 \cos \phi, \qquad (13a)$$

$$W_T = m E I_2 \cos \phi_2. \qquad (13b)$$

The synchronous power is the power corresponding to the torque of the motor acting at synchronous speed, and hence the torque can be readily calculated:

$$\text{Torque in lb.ft. } T = 0.117 \frac{W_T}{n_0}. \tag{14}$$

The output power differs from the synchronous watts because the speed differs from synchronous speed. Also, when the rotor is shorted the difference must be equal to the rotor copper loss:

$$\text{Gross output} \quad W_0 = (1-s) \, W_T \quad \text{in watts.} \tag{15}$$

$$\text{Rotor copper loss } sW_T = mI_2^2 R_2. \tag{16}$$

To obtain the horse-power output the friction loss F must be deducted:

$$\text{H.P. output} = \frac{(1-s) \, W_T - F}{746}. \tag{17}$$

Range of operation

It has been assumed up to this point, that the range of slip variation is between zero and unity, that is, the motoring range. But if the machine is driven mechanically by another source of power while still connected to the A.C. supply, it is possible to have other values of slip. Negative values are obtained by driving the machine above synchronous speed, and positive values greater than unity by driving it in the direction of rotation opposite to that of the field. This does not affect the basis of the equivalent circuit or the equation in any way; what it does affect is the general action of the machine in that mechanical power is supplied instead of being produced.

Equivalent network for induction motor

In the theory presented in this book, which is mainly concerned with machines where the secondary is not short-circuited, an equivalent circuit similar to that of Fig. 5, which includes an ideal induction motor, is generally the most useful. For the ordinary induction motor, however, where the secondary is short-circuited, a useful simplification can be made by eliminating the ideal induction motor as in Fig. 6. This diagram shows the usual form of the equivalent circuit of the induction motor.

It is evident that the current I_2 remains the same if both the voltage and the impedance in the secondary part of the circuit are divided by s, that is, if sE is replaced by E, R_2 by R_2/s, and sX_2 by X_2. If this is done in Fig. 5, the ideal induction motor can be taken away, and there remains the simple network shown in Fig. 6.

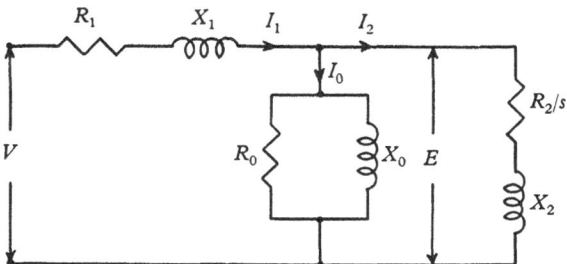

Fig. 6. Normal equivalent circuit of induction motor

It should be noted that in this equivalent circuit, the input to the secondary is $mI_2^2R_2/s$, i.e. $1/s$ times the rotor copper loss. The input to the secondary in the equivalent circuit of Fig. 6 is the sum of secondary copper loss and mechanical output, and is thus a direct measure of the torque in terms of synchronous power.

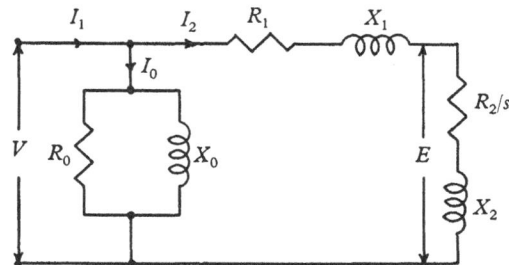

Fig. 7. Simplified equivalent circuit of induction motor

The circuit can be simplified still further by shifting the branch R_0X_0 to the terminals as shown in Fig. 7. This assumes that both magnetizing current and iron loss are constant at all loads. The simplification is not acceptable in the analysis of a main motor, for the error is too large, but it is sufficiently accurate for use in analysing an auxiliary machine, and is used for that purpose in later chapters.

3. Representation of characteristics

Torque-speed curves

The function of any electric motor is to exert the torque required to drive some machinery or apparatus at an appropriate speed. With any particular adjustment of the components of the motor or its controlling apparatus, and with a fixed supply voltage and frequency, the torque exerted at a given speed has a definite value. If the speed varies, the torque exerted also changes. The relation between the torque and the speed is the *torque-speed characteristic*, which can be represented by a curve relating the two quantities.

For most duties a motor having a single torque-speed characteristic is all that is required; where the change of speed with load is small, it is generally known (rather inaccurately) as a *constant-speed motor*. For such duties the ordinary induction motor, either of the slip-ring or squirrel-cage type, is suitable, and has great advantages from the point of view of robustness and simplicity. But for many applications some adjustment of the speed at a given torque is desirable. For such applications, the motor must be capable of operating with different torque-speed characteristics at different times, the adjustment being made by means of a controlling mechanism or device either on the motor or separate from it. Such a motor is known as a *variable-speed motor*, although the term *adjustable-speed motor*, which is sometimes used, is a more accurate description.

The shunt-excited D.C. motor is a simple example of a variable-speed motor. With a given value of field resistance, the torque-speed characteristic is a curve such as curve *A*, Fig. 8, on which the speed, though approximately constant, drops slightly as the torque increases. When the field resistance is reduced, the torque-speed characteristic changes to other similar curves *B*, *C* or *D*, on which the speed at a given torque is lower. In general, the characteristics of a variable-speed motor are expressed by a family of curves, each of which corresponds to a particular setting of the control device.

The type of characteristic obtained with a D.C. shunt motor, where the speed is approximately constant over a range of torques, is called a *shunt characteristic*. By analogy with the D.C. series motor, a characteristic on which the speed varies greatly as the torque varies, is called a *series characteristic*. These two terms are commonly used to describe the torque-speed characteristics of A.C. commutator motors.

Current locus curves

It is also important to know the value of the current taken from the supply under any condition of operation. In an A.C. machine, the current has two components, viz. a power component, and a wattless component. The curve relating the two components of current to each other is known as a *current locus curve*. The well-known 'circle diagram' of the induction motor is an example of such a current locus curve.

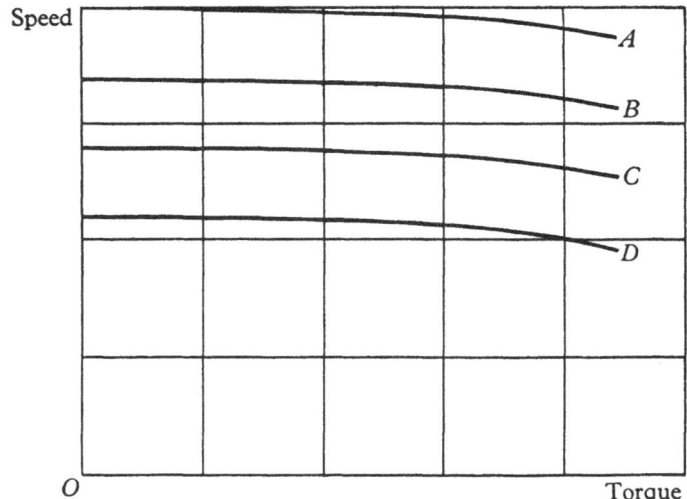

Fig. 8. Family of torque-speed curves of D.C. shunt machine

On the current locus diagram the in-phase or power component I_p is usually plotted as the ordinate, and the quadrature or wattless component I_q as the abscissa, as in Fig. 9. The current corresponding to the point P is then represented both in magnitude and phase by the vector OP, the supply voltage vector V being in the vertical direction. As indicated in Fig. 9, the current lags behind the voltage by the angle ϕ, and the power factor is $\cos \phi$.

On this diagram the real axis is vertical and the horizontal axis to the right is the negative imaginary axis. Thus the diagram is rotated through 90° compared with the usual Argand diagram for complex quantities. This method of plotting agrees with the usual convention for machine diagrams according to which the power component of current is vertical. Therefore, if the complex expression $I_1 = I_p + jI_q$ for the primary current represents a lagging current as shown in Fig. 9, the value of I_q is negative.

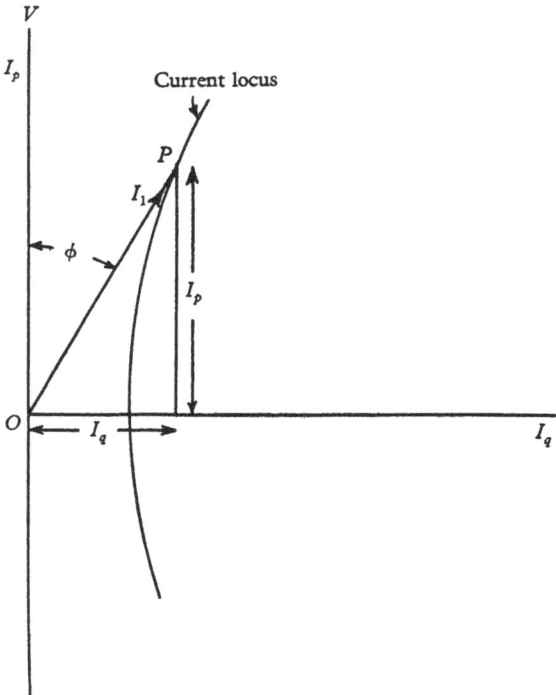

Fig. 9. Current locus diagram

A variable-speed motor has a different current locus curve for each setting of the control device, so that the complete characteristics are given by a family of curves, as with the torque-speed curves. For a given setting, each point on the current locus curve corresponds to a point on the torque-speed curve and to a particular value of speed or slip. To link the two curves, values of slip should be marked at points on the current locus curve.

Thus, for each value of slip there are definite values of torque, in-phase current and quadrature current, and these four quantities tell all that it is required to know about the performance of the machine. The two curves, namely the torque-speed curve relating the two output quantities, and the current locus curve relating the two input quantities, show at a glance the essential relationships which determine the performance of the motor.

Calculation of characteristics

The equations derived for any type of machine can be used to calculate the characteristics. For each setting of the control device, there is a set of constants in the equations, which determine the torque-speed and current locus curves for that adjustment. The equations of the induction motor with short-circuited rotor have already been given, and other equations for the various types of polyphase commutator machine will be derived later. In general, the independent variable is taken as the slip s, and the torque and the two components of current can be calculated for each value of slip.

Other quantities, such as resultant current, power factor, kW. input, kVA. input, H.P. output, and efficiency, follow directly from these four quantities. A great variety of characteristic curves relating any two quantities can be plotted, and are useful for particular purposes, but in a general study, such as that given in this book, the two main characteristic curves are quite the most useful. The study of the characteristics of the various types of commutator machine will be mainly a comparison of their torque-speed and current locus curves.

Circle diagram and torque-speed curve of induction motor

If the vector quantity I_1 is calculated from equation (12) for a number of values of slip s, pairs of values of the two components I_p and I_q are obtained, and the current locus curve can be plotted. It will now be proved that the curve is a circle.

Equation (12) can be expressed in the form

$$I = I_p + jI_q = \frac{(a+sb)+j(c+sd)}{(f+sg)+j(h+sk)} \times V. \tag{18}$$

Hence $\quad (fI_p - hI_q - aV) + s(gI_p - kI_q - bV) = 0,$

and $\quad (hI_p - fI_q - cV) + s(kI_p + gI_q - dV) = 0.$

Elimination of s gives

$$I_p^2 - \frac{(bh+cg-ak-df)\,V}{gh-fk}\,I_p + I_q^2 + \frac{(ck+ag-dh-bf)\,V}{gh-fk}\,I_q$$

$$= \frac{(ad-bc)\,V^2}{gh-fk}. \tag{19}$$

Since the coefficients of I_p^2 and I_q^2 are equal and there are no terms in $I_p I_q$, this is the equation of a circle. In the induction motor, the term h is zero and equation (19) can be simplified for this case by omitting products of h. But for some machines—the Schrage motor, for example—h is not zero, and the complete form of equation (19) is required.

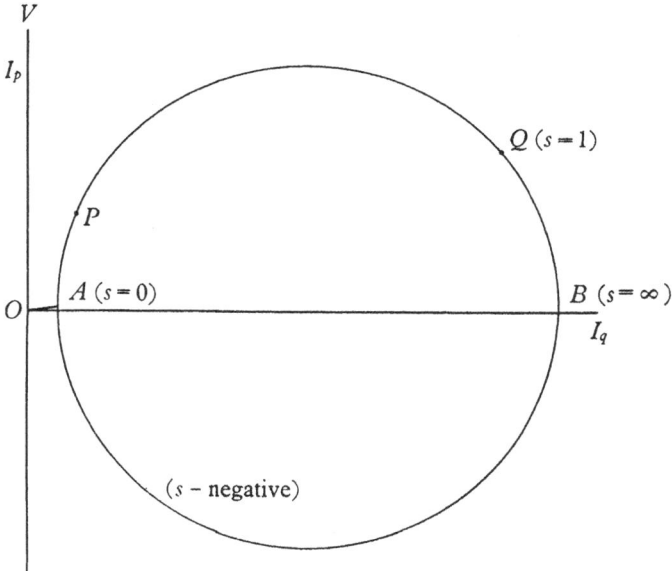

Fig. 10. Circle diagram of induction motor

This method of obtaining the current locus is quite general, and involves only those assumptions applying to the equivalent circuit Fig. 6. Simplification can be introduced into equation (12) to correspond to the simplified equivalent circuit of Fig. 7; such a modification results in the well-known Heyland circle diagram. Thus the general equivalent circuit and the equations derived from it can readily be used as they stand, but, if simplifications are made, the nature of them can be clearly seen.

A typical circle diagram is shown in Fig. 10, which is a current locus curve of the type shown in Fig. 9. A is the point where $s = 0$; hence the no-load current is OA. B corresponds to $s = \infty$ and Q to $s = 1$. At standstill the current OQ is large and at a low power factor. The part of the curve between Q and B corresponds to

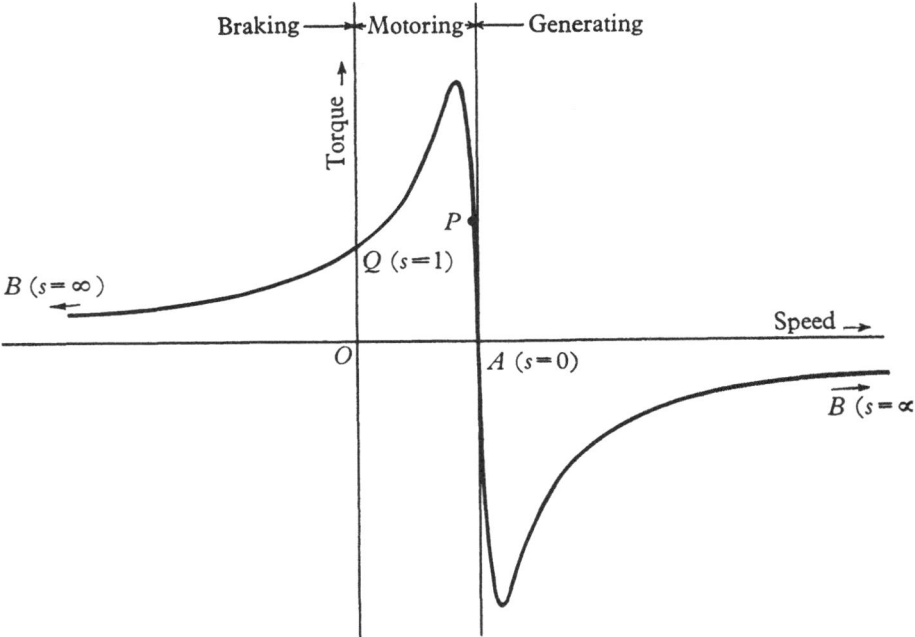

Fig. 11. Torque-speed curve of induction motor

operation of the motor in the reversed rotation, that is, with slips greater than unity. The working range is AP, over which the value of slip is small, and generator operation at super-synchronous speed with negative slip is obtained between A and B on the lower half of the circle. It should be noted that the size and position of the circle does not depend on R_2, which merely affects the value of slip corresponding to each point.

The torque-speed curve can be obtained directly from the equations by calculating the torques for a number of values of slip. In Fig. 11, which shows a typical curve for an induction motor, the same letters are used as in Fig. 10 to indicate the normal load, synchronous, standstill and infinite slip points. The torque reaches a maximum at the 'pull-out' point between P and Q. The shape of the curve in the generating region, where the slip is negative, is similar to that of the part for positive slip. At negative speeds between Q and B, mechanical power is supplied to the shaft and the electrical input is also positive, all the power being dissipated as losses. This part is marked on the curve as the region of braking operation.

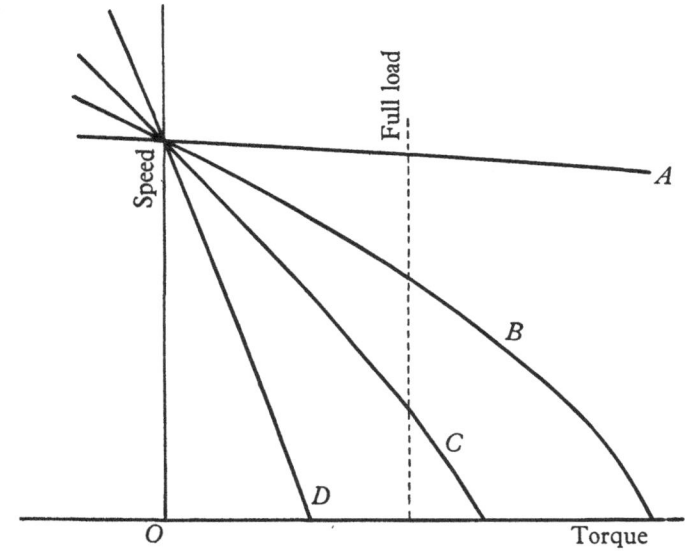

Fig. 12. Induction motor torque-speed curves showing the effect of increased secondary resistance

When the main concern is with the working range as a motor, it is more convenient to plot the torque as abscissa and the speed as ordinate as in Fig. 8, and this method of plotting is used in later chapters for the study of variable-speed commutator motors. In Fig. 12, curve A shows part of the torque-speed curve of an induction motor with the secondary shorted, and curves B, C and D are the curves obtained when the secondary resistance is increased, for example, by connecting external resistance in the rotor circuit of a slip-ring motor. If s and R_2 are increased in the same ratio in the equations, the same currents flow. Hence curves B, C and D can be obtained from curve A by increasing the slip at each point in the ratio by which R_2 is increased. The insertion of rotor resistance is commonly used for limiting the current and increasing the torque when starting slip-ring induction motors, and it may also be used for varying the speed when running. It provides a simple example of the use of an 'injected voltage', to regulate the speed of the motor. The injected voltage is the drop in the external resistance.

The method of calculation described above, by which the current and torque are obtained by substituting in a vector equation, is much the most straightforward and the easiest for practical work both for induction motors and commutator machines. Many ingenious geometrical constructions have been devised for particular combinations, and papers have been written to prove that the current locus curves of particular types of machines are circles, but for many practical cases, these methods are either very involved or not possible at all. The analytical method given in this book is a perfectly general one, and can be used for any machine system, however complicated. It has been used with good results for a wide variety of practical commutator machines.

4. Operation of the induction motor with an injected voltage

Vector diagrams and equivalent circuit

It has already been explained that the normal induction motor, with its rotor short-circuited, operates at a speed just below synchronous speed, and at a lagging power factor. But this is only a special condition. If the secondary is not short-circuited, but is connected to a separate source of voltage, the characteristics of the motor can be modified so as to obtain speed variation or power-factor correction or both. The voltage, which may be obtained from a commutator machine or other external apparatus, must have the same frequency, number of phases, and phase sequence as the voltage induced in the secondary. This external voltage applied to the secondary terminals is called an *injected voltage*, and is denoted by E_k.

The value of the injected voltage depends on the way in which it is produced, that is, on the auxiliary apparatus used and on the way in which this apparatus is connected. The voltage may be constant, or may vary with the load on the motor or with its speed. The study of polyphase commutator machines is basically a study of the various methods of producing the injected voltage and of the consequent effect on the motor characteristics.

Vector diagrams and equivalent circuit with an injected voltage

The injected voltage modifies the vector diagrams of Figs. 2, 3 and 4 to those of Figs. 13, 14 and 15. The essential change is that a voltage E_k has been added to the diagram of referred secondary voltages, Fig. 15. In Fig. 15, it is the vector difference of sE and E_k which equals the secondary impedance drop. In this diagram, E_k is not the actual value of the injected voltage but the *referred injected voltage*, obtained by multiplying the actual value by the transformation ratio K. Nevertheless, in the subsequent theory, E_k is called simply the 'injected voltage' in line with the other secondary quantities.

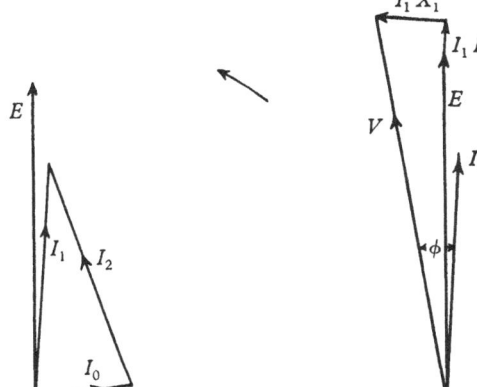

Fig. 13. Primary currents
with injected voltage

Fig. 14. Primary voltages with
injected voltage in secondary

In Fig. 15, the introduction of E_k into the secondary circuit in the directions shown has two results. In the first place, the slip and the induced voltage sE are increased. Secondly, the secondary current I_2 is advanced in phase compared with its value in Fig. 4. Consequently, since the magnetizing current I_0 is practically unchanged, the primary current I_1 is also advanced in phase compared with its value in Fig. 2 as can be seen in Fig. 13. Thus the angle ϕ

between V and I_1 is reduced: in other words, the power factor of the motor is improved. The primary voltage diagram is modified as shown in Fig. 14.

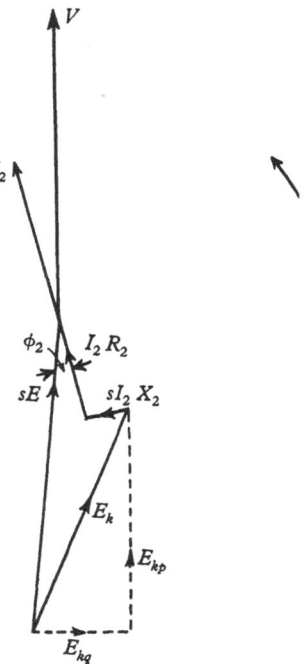

As before, Fig. 15 may be considered to be a diagram of the slip-frequency voltages in the secondary circuit of the equivalent unity ratio motor. Alternatively, it may be considered as a diagram of supply frequency voltages in the primary circuit, each corresponding to an actual secondary voltage. The former conception is useful when considering the actual conditions in the secondary circuit. The latter conception, according to which all secondary quantities are referred to the primary frequency, is used in developing the general theory. In this conception, E_k has a definite magnitude and phase angle relative to the supply voltage V, and can be resolved into two components, E_{kp} in phase with V, and E_{kq} in quadrature with V. If the vector E_k is such as to correct the power factor, the component E_{kq} is negative.

Fig. 15. Secondary voltages with injected voltage

The equivalent circuit of Fig. 5 is modified to allow for the effect of the injected voltage, by introducing the voltage E_k on the secondary side of the ideal motor, as shown in Fig. 16. This circuit corresponds to the vector diagrams of Figs. 13, 14 and 15, and forms the basis of the theory used in the remainder of the book for determining the performance of commutator motors, or of induction motors with commutator machines connected in the rotor circuit.

Equations of the induction motor with an injected voltage

Equations (7) and (8) still hold for the modified equivalent circuit of Fig. 16, except that equation (8 d) now becomes

$$sE = I_2 Z_2 + E_k. \tag{20}$$

Equations (9) and (12) are unchanged, but the expression for I_2 is modified by the introduction of E_k. Thus

$$V = cE + I_2 Z_1, \qquad (9)$$

$$I_1 = \frac{VY_0 + I_2}{c}, \qquad (12)$$

$$I_2 = \frac{sV - cE_k}{sZ_1 + cZ_2}. \qquad (21)$$

Further equations giving the value of E_k or relating it to the other quantities, are necessary to determine the characteristics of the motor. These expressions depend on the particular arrangement of auxiliary apparatus, and several examples are given in later chapters for the principal types of commutator machine. When these relationships are known, the current locus curve and the torque-speed curve can be calculated as explained in the previous section.

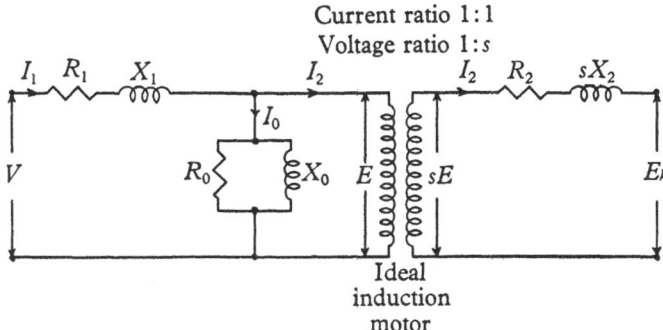

Fig. 16. Equivalent circuit of induction motor with injected voltage

If, for a given value of slip, the injected voltage E_k is known in magnitude and phase, this completely determines the input current and torque of the motor. Hence the operation of the motor can be completely controlled by controlling the injected voltage. Conversely, if a particular condition of operation, defined by torque, power factor, and speed, is required, there is always a value of injected voltage which will produce this condition, and it can be calculated by means of the equations.

The power relationships given by equations (13) to (17) are affected by the introduction of the injected voltage, because power is taken from or passed into the secondary of the motor by the external apparatus. This power is

$$W_k = mE_kI_2 \cos \phi_k, \tag{22}$$

where ϕ_k is the angle between E_k and I_2.

The power sW_T passed to the secondary winding now equals the sum of the secondary copper loss and the power W_k, and equation (16) becomes

$$sW_T = mI_2^2R_2 + W_k. \tag{23}$$

The synchronous power is still given by

$$W_T = mEI_2 \cos \phi_2 \tag{13b}$$

and the output torque, output watts, and horse-power by

$$T = 0.117 W_T/n_0, \tag{14}$$

$$W_0 = (1 - s) W_T, \tag{15}$$

$$\text{H.P.} = \frac{(1 - s) W_T - F}{746}. \tag{17}$$

As a general rule, if the main induction machine is running as a motor below synchronous speed, electrical power is taken from the rotor by the regulating apparatus. The simplest example of the regulation of an induction motor by an injected voltage is the use of an external secondary resistance as mentioned on p. 20. The injected voltage is here the ohmic drop in the resistance, and the power W_k is the loss in the resistance. When a commutator machine or other apparatus is used to provide the injected voltage, the power is recovered, apart from relatively small losses in the regulating apparatus. The resulting increased overall efficiency constitutes one of the important advantages obtained by using commutator machines.

Effect of the injected voltage on the characteristics

If the torque exerted by the motor remains constant, the effect of applying an injected voltage is in general to alter both the speed and the power factor of the motor. The result can be seen most clearly by considering first the condition when the motor is unloaded, corresponding to the vector diagrams of voltages and currents shown in Fig. 17.

As in Figs. 13–15, the magnitude and phase of the injected voltage in Fig. 17 are such as to bring about a reduction of speed and an improvement in the power factor. The three vector diagrams corresponding to Figs. 13–15 are there shown for the no-load condition on a single figure. The primary voltage V is drawn vertically, and the two components of E_k, namely E_{kp} in phase with V and E_{kq} in quadrature with V, are indicated. E and V are almost in phase, and the secondary current I_2 is approximately horizontal for the unloaded condition. The wattless component of I_2 depends mainly on the quadrature component E_{kq}, and as the wattless component of I_2 increases, so that of I_1 decreases.

It is evident from the diagram that the following two rules are approximately true:

(1) A component of injected voltage in phase with the primary voltage produces speed variation without greatly affecting the power factor.

(2) A component of injected voltage in quadrature with the primary voltage modifies the power factor without greatly affecting the speed.

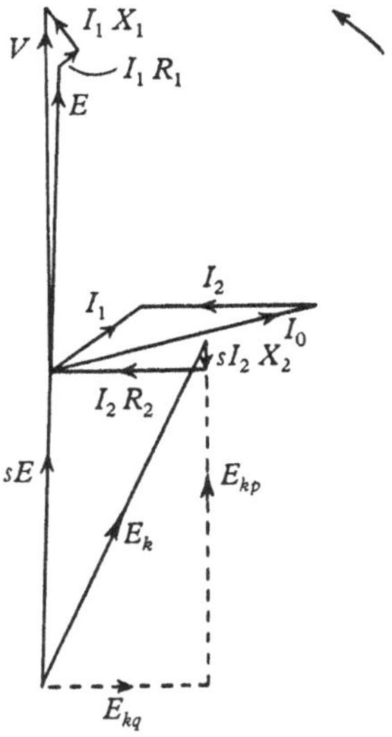

When the motor is loaded, the rules are still approximately true, although the

Fig. 17. Current and voltage vectors on no-load with injected voltage

accuracy is less than for the no-load condition, because of the primary and secondary impedance drops. The rules are in any case much too inaccurate for calculation purposes, but they do give a picture of the general way in which the characteristics of the motor are controlled by the injected voltage.

Either of the components of E_k can be reversed. If the in-phase component E_{kp} becomes negative, the speed of the motor rises above that obtained with the secondary shorted, and, if the value is sufficient, the speed is above synchronous speed. On the other hand, if the quadrature component becomes negative, the power factor becomes more, instead of less, lagging. Hence, with a suitable control of the injected voltage, speed variation is possible above synchronous speed as well as below, and any desired value of power factor can be obtained at any speed.

A commutator machine used to supply an injected voltage for the purpose of varying the speed of an induction motor is called a *speed-regulating machine*. Generally, such a machine improves the power factor at the same time. A commutator machine used to supply an injected voltage, in order to correct the power factor without varying the speed, is called a *phase advancer*.

Injected voltage in phase with the secondary current

If a more accurate determination of the effect of a given injected voltage than that given by the two approximate rules on p. 25 is required, a calculation can be made using equations (12) and (21). There is, however, one condition when the exact result can be seen immediately. This arises if the injected voltage is in phase with the secondary current, and it occurs in practice when an external secondary resistance is used, as well as with certain types of commutator machines. Fig. 18 is a diagram of secondary voltages for this condition. The angle between I_2 and sE is the same as it would be if there were no injected voltage, because, if E and I_2 remain unchanged, the vectors sE and sI_2X_2 increase in the same ratio when the slip is increased. Thus, if the injected voltage is in phase with the secondary current, the power factor at a given torque is exactly the same as if there was no speed variation.

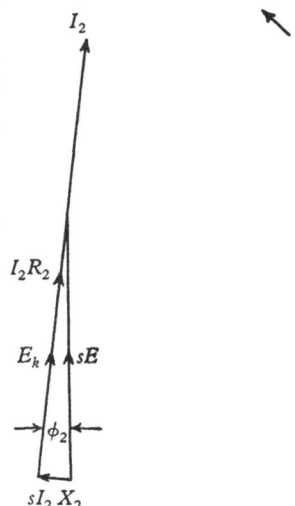

Fig. 18. Vector diagram with injected volts in phase with secondary current

This result does not agree exactly with the first rule on p. 25 because there is always a phase angle between I_2 and V, and consequently a small quadrature component of E_k. The example shows, in fact, the nature of the error, which is due mainly to the primary and secondary impedance drops. If the quadrature component E_{kq} were changed without altering E_{kp}, the main effect would be to modify the power factor, but there would also be a small change of speed. Thus, the second rule, like the first, is only an approximate one. The rules are very useful, however, because they show clearly the fundamental idea underlying the speed variation of induction motors, namely, that if an injected voltage is applied, the secondary induced voltage, and hence the slip, must adjust themselves so as to obtain equilibrium.

Transition from sub-synchronous to super-synchronous speed

If the motor operates with a given primary current at a given power factor the secondary current and the torque are thereby determined irrespective of the speed. If the speed changes, the value of the vector E_k required to maintain equilibrium changes correspondingly. It is convenient to consider the effect of changes of E_k when the primary current and the power factor remain constant.

If the in-phase component of the injected voltage is increased gradually from zero in a negative direction so as to raise the motor speed while the torque remains the same, the slip first decreases to zero and then changes sign. If the secondary phases are perfectly balanced, the transition through synchronism is continuous, apart from the negligible hysteretic torque mentioned on p. 7. Thus the synchronous condition, when the secondary currents are direct current, is only the limiting case of zero frequency alternating currents in the secondary circuit. Below synchronism the currents in the various secondary phases rise and fall at different times but in a regular sequence. At synchronous speed the currents assume fixed values; above synchronous speed they vary in regular sequence again, but the order through the phases, called the *phase sequence*, is reversed. The reversal of phase sequence occurs because the relative motion between the rotating flux wave and the rotor is reversed.

In this chapter the method is first of all to establish vector diagrams relating all the currents and voltages in the motor, and then, by the use of suitable conventions, to derive from the diagrams an equivalent circuit and equations expressing the relations. In considering operation at super-synchronous speeds the first thing is to study the vector diagrams. An example of the secondary voltage diagram at a speed above synchronism for a motor with unity ratio is shown in Fig. 19. The torque and power factor are assumed to remain as in Figs. 15, and the primary diagrams of Figs. 13 and 14 still apply.

In order to link up the diagram of Fig. 19 with the primary diagrams of Figs. 13 and 14, the phase angle relation between I_2 and sE in Fig. 19 must be as shown. The direction of the vector E remains vertically upwards, but sE is reversed. The angular relation in space between the rotating flux wave and the rotating wave of secondary M.M.F. corresponding to Fig. 19 is the same as for the sub-synchronous condition to which Fig. 15 applies. However, owing to the change in the relative rotation between flux and rotor, the slip-frequency current I_2 lags behind sE, instead of leading as it does at sub-synchronous speeds. This difference can be allowed for by reversing the direction of rotation of the vectors in Fig. 19, as indicated by the arrow.

The slip s has a negative value, and the frequency, which cannot be negative, is now $-sf$. The reactance drop must also be positive, and is therefore $-sI_2X_2$ (a positive quantity) leading I_2. Although the diagram applies to one phase only, the change in the direction of rotation of the secondary vector

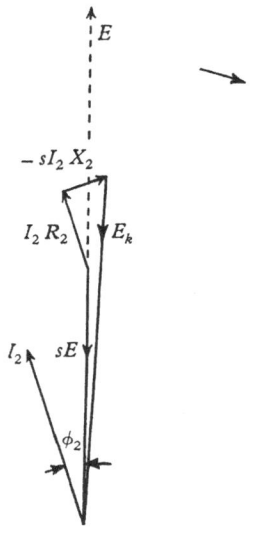

Fig. 19. Slip-frequency secondary voltages at super-synchronous speed

diagram above synchronous speed is associated with a reversal of the phase sequence in the three (or more) secondary phases. It can be said that the reversal of the arrow in Fig. 19 is caused by the change of phase sequence, because this change expresses the fact that the relative motion between flux and rotor is reversed.

In order to obtain a vector diagram of referred secondary voltages at supply frequency as mentioned on pp. 9 and 22, it is only necessary to change the direction of the arrow to indicate counter-clockwise rotation of the vectors as in Fig. 20. All voltages remain unchanged except that the vector for the reactance drop now lags behind the current. It is therefore necessary to introduce the fiction of a negative reactance drop sI_2X_2 corresponding to the negative slip s. Using Fig. 20, there is now a definite phase angle between E_k and the primary voltage V, and E_k can be resolved into components E_{kp} and E_{kq} as before.

Characteristics of an induction motor with a constant injected voltage

Many practical cases approximate to the condition where the injected voltage E_k is of constant magnitude and at a constant phase angle to V. When this condition is fulfilled, the performance can be calculated from the equations of pp. 22–4 using a constant value for the vector E_k. Definite values of the two components E_{kp} and E_{kq} result in definite torque-speed and current locus curves, but a change in either component or both, changes both of the characteristic curves. By the method given on p. 16 it can be proved that when the vector E_k is constant, the current locus curve is always a circle.

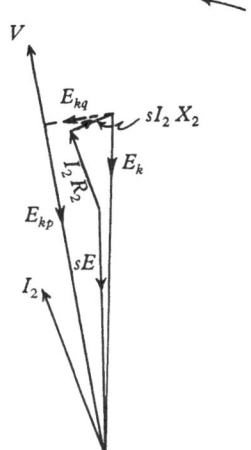

Fig. 20. Supply frequency referred secondary voltages at super-synchronous speed

The effect of a constant injected voltage on the torque-speed curve is to raise or lower the characteristic above or below synchronous speed, while the change in speed from no-load to full-load, although increasing slightly at the lower speeds, still remains small. Thus the motor becomes a variable-speed motor with shunt characteristics similar to those of a D.C. shunt motor. There is, however, no limitation to the speed range available, as there is in the D.C. shunt motor, and the family of torque-speed curves can extend right down to standstill as shown in Fig. 21. The torque-speed curves are still of the same type when, in order to give power-factor correction, a quadrature component E_{kq} is used in combination with E_{kp}.

The current locus curves depend both on the quadrature component E_{kq} and the in-phase component E_{kp}. Fig. 22 shows the circle diagrams obtained when $E_{kq} = 0$ with various values of E_{kp}. When $E_{kp} = 0$, the current locus curve AP is the ordinary circle diagram of the induction motor shown in Fig. 10. At the sub-synchronous speeds obtained when E_{kp} is positive, however,

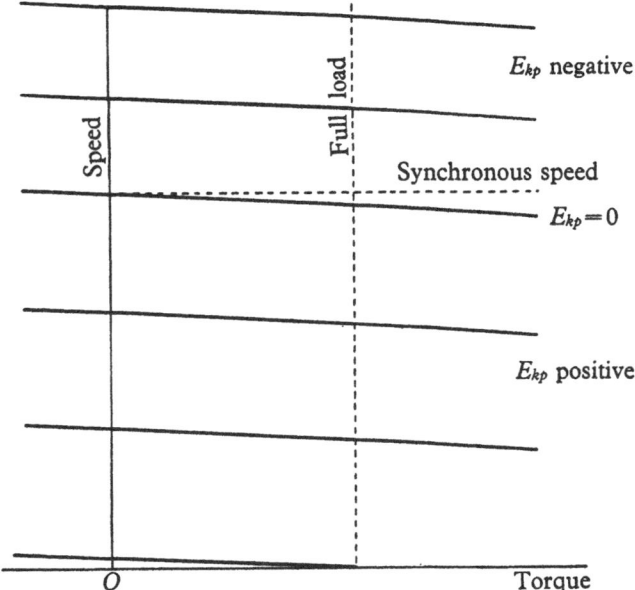

Fig. 21. Torque-speed curves with injected voltages

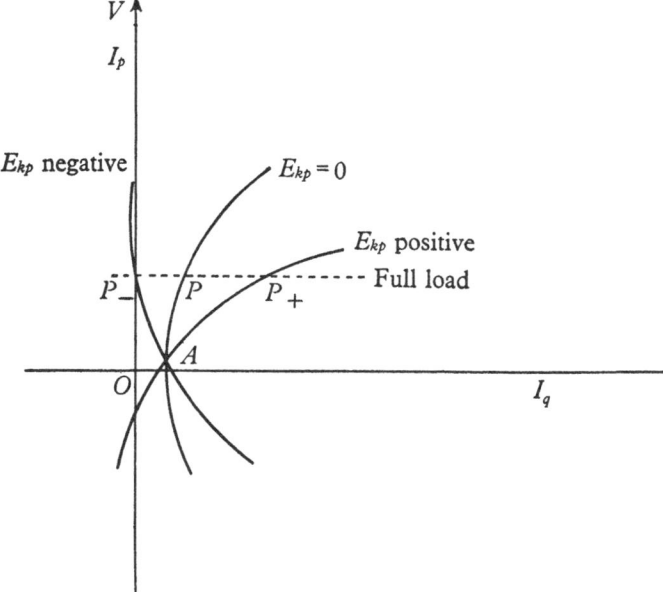

Fig. 22. Circle diagrams with injected voltages, E_{kq} zero

the secondary reactance drop sI_2X_2 increases and causes the current I_2 to lag more. As a result, the current locus curve $AP+$ is tilted over to the right as shown. On the other hand, above synchronous speed the direction of the vector sI_2X_2 reverses, causing the current I_2 to lead the induced voltage. The current locus curve above synchronous speed is therefore the circle $AP-$, the centre of which is above the horizontal axis. The point of intersection A is approximately the no-load point of the induction motor operating alone.

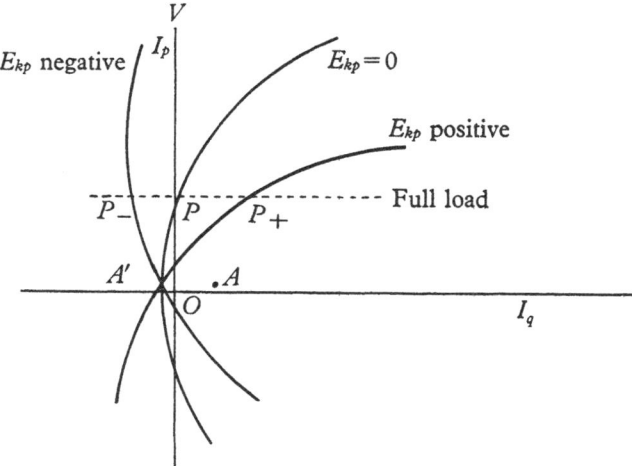

Fig. 23. Circle diagrams with injected voltages, E_{kq} negative

The main effect of a quadrature component E_{kq} on any of the current locus curves is to move the centre of the circle to the left or right. The curves in Fig. 23 apply for different values of E_{kp} with a common negative value of E_{kq}, which has the effect of correcting the power factor at all speeds. The point of intersection changes from A to A', but the shape of each characteristic is similar to that of the corresponding curve of Fig. 22. It is evident from the circles that if the same power factor is required at a given torque over a range of speeds, the value of the quadrature component E_{kq} must be greater at the lower speeds.

Figs. 22 and 23 show the current taken from the supply by the induction motor primary winding. They do not indicate the overall current of the motor and its regulating apparatus; the current taken from the supply by the regulating apparatus must be combined with that of the motor primary to obtain the overall current locus curve. At super-synchronous speeds the magnitude of the total current is greater than that of the primary current, while at sub-synchronous speeds it is generally less.

It has been assumed in the above description that the voltage E_k at the secondary terminals is constant. In a practical equipment, however, the regulating apparatus will have its own impedance which will cause E_k to vary slightly with the load. This condition can be dealt with by including the impedance of the external apparatus with the secondary impedance of the motor, and applying a constant value of E_k to the circuit so obtained.

Chapter 2

THE WINDINGS OF A POLYPHASE COMMUTATOR MACHINE

5. Types of winding

The essential part of a D.C. machine is the armature winding connected to a commutator and rotating in a *stationary* field. In the same way, the most important part of a polyphase commutator machine is a similar commutator winding rotating in a *rotating* field. The rotating field is set up by polyphase currents flowing either in other windings or in the armature itself.

The principal difference in construction between a D.C. machine and an A.C. commutator machine is due to the fact that the magnetic circuit of the A.C. machine has to be laminated throughout, on account of the alternation of the flux. For this reason, the stator of an A.C. commutator machine is similar to that of an induction motor, and consists of a laminated core carrying coils in slots.

The windings used in polyphase commutator machines can be classified under the two headings:

(1) Armature winding. (Commutator winding.)

(2) Other windings.

In this chapter the properties of these windings are discussed in preparation for the next chapter which deals with the way in which the windings are combined together to form the various types of machine.

Fundamental properties of polyphase windings

The study of the operation of any polyphase machine is based on three fundamental properties of the windings.

(1) A rotating flux induces a polyphase system of voltages in the winding.

(2) When a polyphase system of currents flows in the winding, it sets up a rotating flux.

(3) A rotating flux acting on a polyphase system of currents of the correct frequency produces a steady torque.

It is usual to concentrate on the voltage or current in one phase and to use this to represent the polyphase system.

Armature winding

The armature winding itself is a double-layer drum winding connected to a commutator and is essentially the same as the armature winding of a D.C. machine. The brush gear, however, is arranged so as to collect polyphase currents from the commutator, and has a set of brushes for each phase.

The winding may be of the wave or lap type, as in a D.C. machine, depending on the number of turns in series required between adjacent segments. If necessary, parallel circuits can be employed giving 'series-parallel' windings in the wave type and 'duplex' windings in the lap type. Since these windings are fully described in books on D.C. machines it is only necessary to mention those special points peculiar to A.C. machines.

Wave windings are used on small machines where it is possible to have more than one turn in series between adjacent segments. A wave winding has the advantage that it is not necessary to provide equalizing connectors between equipotential points on the commutator. With a lap winding it is essential to provide equalizers, which are usually located at the inner end of the commutator.

The commonest type of armature winding on medium-sized and large commutator machines is a simplex lap winding with one turn per coil. A developed diagram for a winding of this type is shown in Fig. 24. For reasons which will be discussed in Chapter 4, the flux per pole which can be carried with this type of winding is limited, unless interpoles are provided. Therefore, large machines without interpoles often have duplex windings, in which the ends of a turn are connected to alternate segments while the intermediate segments are connected to another similar set of coils entirely separate from the first set. In addition to the equalizer connexions at the back of the commutator, a duplex winding must also be provided with some means of maintaining the correct distribution of potential between the two halves of the winding.

Fig. 25 shows one method of doing this. It is a developed diagram of a duplex lap winding with equalizing connectors provided between the two halves of the winding in the following manner. Each coil of one half-winding has an equalizer connected to its mid-point at the non-commutator end of the machine. The other end of the equalizer is connected to the segment that lies between those to which the coil ends are connected. These

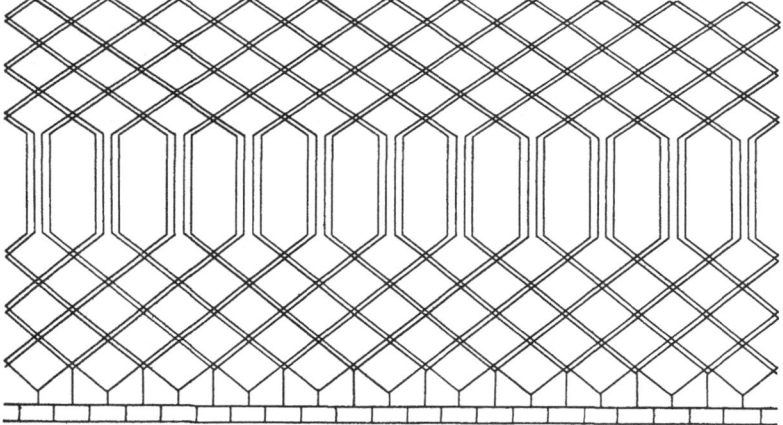

Fig. 24. Simplex lap winding

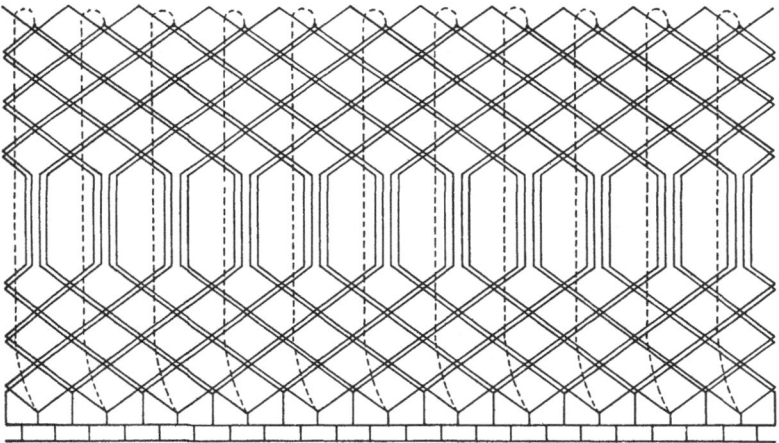

Fig. 25. Duplex winding with back-to-front connectors

equalizers are therefore called 'back-to-front connectors' and they pass from one end of the armature to the other through spaces between the core and the shaft.

Another method of equalizing the two halves of a duplex winding is described on p. 83.

Other windings

The other active windings of a polyphase commutator machine are distributed in slots round the periphery of the core, and may be on the stator with direct connexion to stator terminals, or on the rotor with connexions through slip-rings. The rotor core is always uniformly slotted, and so is the stator except when interpoles are provided. A distributed winding on a uniformly slotted core is exactly like that of an induction motor, except that numbers of phases greater than three are often used for the secondary windings of commutator machines.

When interpoles are used, the arrangement of the stator windings has to be modified as explained in Section 16, but the essential operation is the same. For purposes of explanation the winding can be considered to be equivalent to a distributed polyphase winding which sets up a rotating flux. The provision of interpoles, as well as the addition of interpole windings or damping windings in order to assist commutation, is a matter of detail, and does not affect the general operation of the machine.

Subject to this qualification, all polyphase commutator machines are built up by combining together an armature winding, and one or more A.C. windings of the type used on induction motors. The latter windings may be located on either stator or rotor.

6. The induced voltage in the armature

The arrangement of the armature and brush gear of a typical three-phase two-pole machine is shown diagrammatically in Fig. 26. The brushes A, B and C are spaced 120 electrical degrees apart, instead of at 180° as in a D.C. machine, and if the armature forms part of a magnetic circuit round the air-gap of which a rotating field is set up, the induced voltages, E_{BC}, E_{CA} and E_{AB} are alternating voltages forming a three-phase system; that is, the phases of successive voltages are 120° apart. The vectors representing the voltages are shown in the triangle in Fig. 27.

As in a D.C. machine, each brush shown in the diagram represents a row of brushes, which are carried either in brush boxes fixed to a stud, or in a brush holder, and are all connected in parallel. In a machine having more than two poles, that is, more than one complete flux wave round the circumference, there is normally one brush stud per phase for each pair of poles. For example, a three-phase six-pole commutator winding has a total of nine brush studs, of which three are connected in parallel in

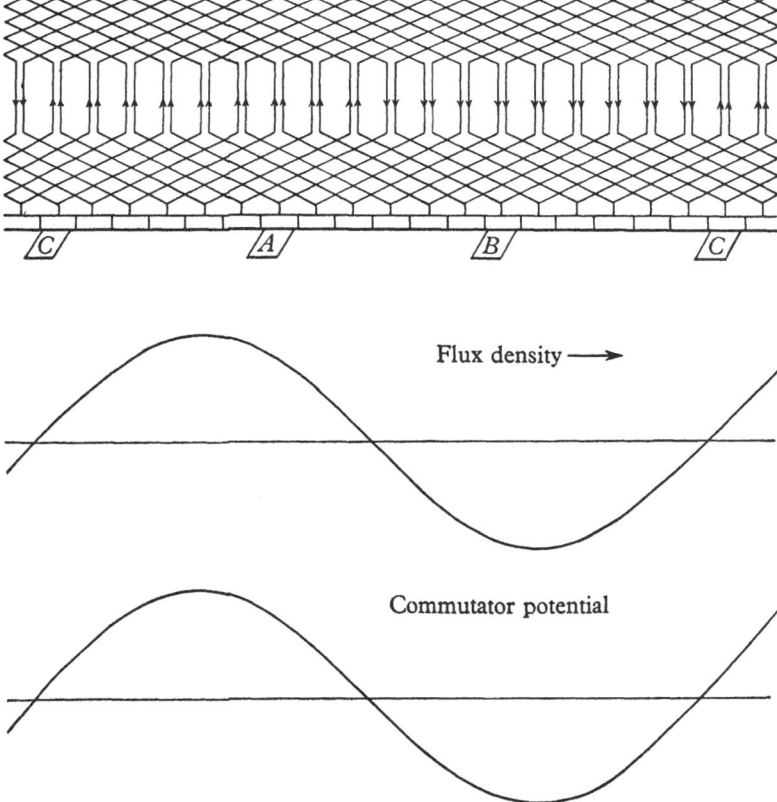

Fig. 26. Distribution of potential round commutator

each phase. In special cases some of the studs in each phase may be omitted. In the diagrams all the brushes connected together in one circuit are indicated by a single brush. In order to make the argument general for machines with any number of poles, two pole pitches only of the machine are considered, and the distances round the circumference are expressed in electrical degrees corresponding to the sinusoidal wave of flux density.

Magnitude and frequency of induced voltage

For a given magnitude of air-gap flux both the magnitude and the frequency of the voltage induced in any particular conductor in a winding is directly proportional to the relative speed of the winding to the field. This is true irrespective of the type of winding.

In an armature winding, the magnitude of the total voltage in the set of turns in series between a pair of brushes is still proportional to the relative speed of the armature and the field, although the set of turns is constantly changing. Thus the presence of the commutator does not affect the *magnitude* of the voltage. But the action of the commutator is such that the frequency of the voltage between a pair of brushes is modified, for, instead of being proportional to the relative speed of the field and the armature, as the conductor voltage is, it is proportional to the relative speed of the field and the brushes. In other words, it is proportional to the speed of the field in space and is independent of armature speed.

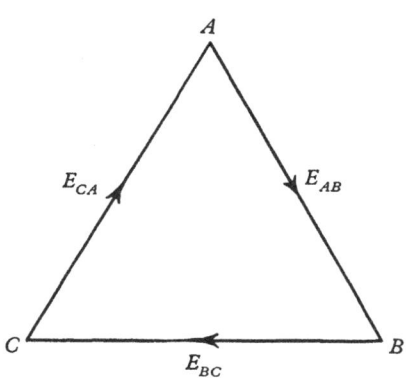

Fig. 27. Triangle of armature voltages

The output voltage from the armature winding of a polyphase commutator machine thus obeys the following two important laws.

(1) The magnitude of the voltage is proportional to the relative speed between the rotating field and the armature.

(2) The frequency of the voltage is proportional to the speed of the rotating field in space.

A D.C. machine can be considered as a special case in which the field is stationary and the frequency is consequently zero. If a low frequency output is required, the field must rotate at a low speed, but the armature may still run at a high speed. Thus a commutator machine is well adapted to provide the variable frequency polyphase voltage required for regulating the speed and power factor of an induction motor.

The reason for these various facts can be seen more clearly from Fig. 26. At the top is shown a developed winding diagram of two pole pitches of a lap-wound armature winding with 100% pitch. The brushes are spaced 120 electrical degrees apart round the commutator. In the centre of the diagram is shown the instantaneous curve of flux density round the machine, which is assumed to be a sine curve moving in the direction of the arrow.

At the bottom of the diagram is shown the curve of distribution of potential round the commutator. The voltage induced in each conductor is proportional to the flux density at the point on the diagram occupied by the conductor, and is indicated by an arrow. The commutator potential is obtained by integrating the voltage per coil round the machine, and is therefore proportional to the integral of the flux density curve with respect to distance round the machine, that is, to another sine curve. The curve of potential in the conductors themselves is therefore displaced 90 electrical degrees from the flux curve, but, owing to the fact that, with the type of winding indicated, the segment to which any conductor is connected is situated 90 electrical degrees away from the conductor, the actual position of the commutator potential curve is the same as that of the flux density curve. Thus, the potential at any point is proportional to the value of the flux density, and if the flux rotates round the air-gap, the potential curve also rotates at the same speed.

The instantaneous voltage between any two brushes is equal to the algebraic difference between the ordinates of the commutator potential curve at the points where the brushes are situated. If the potential curve moves round the air-gap, the voltages between the brushes alternate at a frequency which depends on the speed of rotation of that curve in space. The important result is thus obtained that the frequency of the brush voltage depends only on the speed of rotation of the field in space, whatever the speed of the armature may be. The phase angles between the voltages E_{BC}, E_{CA} and E_{AB} depend on the relative positions of the brushes, and are therefore 120° in the case considered.

Although Fig. 26 shows a full-pitch three-phase lap-wound armature winding in which the opposite sides of a coil are situated at points where the flux densities are equal and opposite, the same result is obtained if the coil pitch is shortened, if there are more than three phases, or if a wave winding is used. In every case, the potential curve travels round the air-gap at the same speed as the flux, and the frequency of the induced voltage is determined by the speed of the flux.

The result may also be deduced from the fact that the voltage between any two brushes at any instant is the sum of the voltages in all the conductors connected between those brushes. This voltage does not depend on which particular conductors are connected between the brushes at that instant, but only on the position of the flux curve. Between two successive instants, the conductors between the brushes change owing to the rotation of the armature, but if the flux does not change, the voltage will remain the same. If the flux wave rotates, the voltage will alternate at the corresponding frequency.

Calculation of induced voltage

The voltage induced in the winding may be calculated in the same way as that in an induction motor winding situated in a rotating field. The virtual value E_{BC} of the voltage between two brushes is the vector sum of the voltages induced in the conductors in series between the brushes, and the fact that the actual conductors are continually changing makes no difference. Hence, as in an induction motor,

$$E_{BC} = \pi \sqrt{2} \Phi(n - n_0) \, (p/2) \, T_a \times 10^{-8}, \tag{24}$$

where Φ = total flux per pole,

n = speed of rotation of armature in rev./sec.,

n_0 = speed of rotation of the flux,

$(n - n_0)$ = speed of rotation of armature relative to the flux,

p = number of poles,

T_a = number of effective turns in series between the brushes.

The equation holds whatever the speed of rotation of the field relative to the armature may be, and whatever its direction. The case of n greater than n_0 corresponds to that of a machine in which a low frequency voltage is generated by rotating the armature at a high speed.

As explained on p. 5, the effective number of turns of a winding is the actual number of turns multiplied by the winding coefficient. The winding coefficient depends on the way in which the conductors of one phase band are distributed round the armature surface. It is clearly an advantage to use a small spread whenever possible, and for this reason the normal three-phase windings used in A.C. machines are arranged with a spread of 60 electrical degrees. In the three-phase armature winding shown in Fig. 26, however, the coils between two brushes constituting one phase are distributed over 120 electrical degrees, and hence the winding is utilized less effectively than is a normal three-phase winding.

Equivalent star connexion of three-phase winding

The three sections *AB*, *BC* and *CA* of the winding in Fig. 26, are of necessity connected in delta at the brushes. The line value of the output voltage is therefore equal to the voltage across each section of the winding, while in a two-pole machine, the line current carried by the brushes is $\sqrt{3}$ times the current in the coils of the winding. In a multipolar lap-wound armature the total line current is divided between a number of parallel paths equal to the number of pairs of poles, but the phases are connected in delta as in the two-pole machine.

When the armature winding is considered in relation to an external circuit, it is more convenient, for the purpose of analysis, to substitute for the delta-connected armature an equivalent star-connected winding, which has the same output voltage and current. With the equivalent star connexion, the phase voltage between a brush and a hypothetical star-point is equal to $1/\sqrt{3}$ times the voltage between brushes, while the current per phase is equal to the actual line current. In Fig. 28 (*a*) the line voltages are represented by E_{BC}, E_{CA} and E_{AB}, while the hypothetical phase voltages in star connexion are E_{OA}, E_{OB} and E_{OC}. In Fig. 28 (*b*) the actual phase currents in delta are I_{BC}, I_{CA} and I_{AB}, but the phase currents in the equivalent star connexion are the brush currents I_A, I_B and I_C. Fig. 28 (*c*) is the vector diagram which gives the relations between the currents.

When the commutator winding is connected to other windings or apparatus, it is simplest to work out the theory in terms of one phase of the complete circuit. In the vector diagrams relating to commutator machines given in later sections, the voltage per phase and current per phase of the commutator winding are those of the equivalent star-connected winding. If E is the phase voltage, equation (24) now becomes

$$E = \pi\sqrt{2}\,\Phi(n - n_0)\,(p/2)\,T_e \times 10^{-8}, \qquad (25)$$

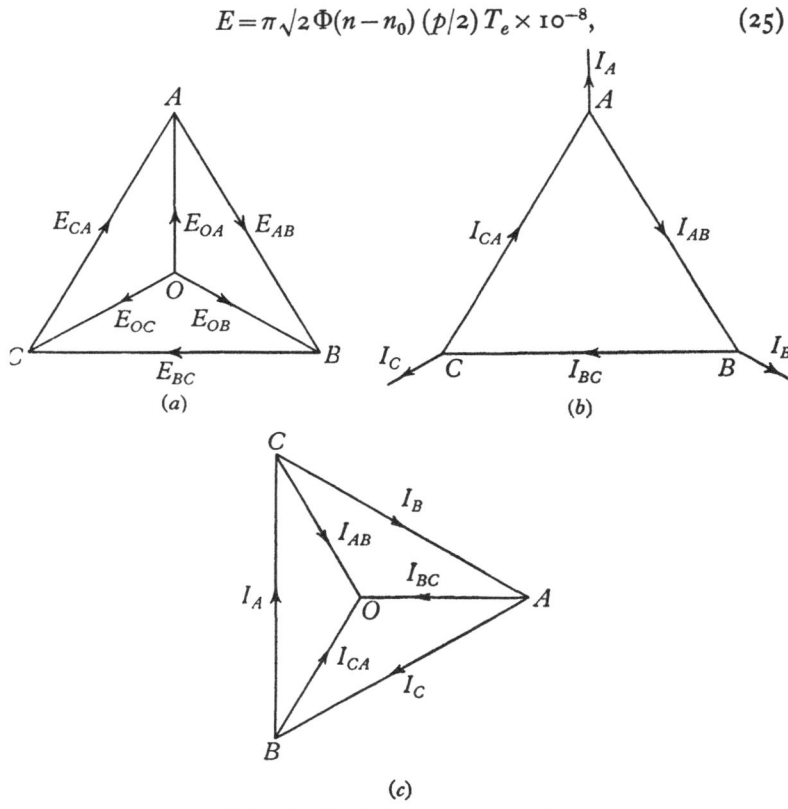

Fig. 28. Equivalent star connexion

where T_e is the effective number of turns per phase in star connexion, that is, the number of turns in an equivalent star-connected winding in which the winding coefficient is unity. It is evident that

$$T_e = T_a/\sqrt{3}. \qquad (26)$$

The brush gear of a polyphase commutator machine is frequently arranged so that the number of phases is greater than three. As an example, Fig. 29 is a vector diagram of the voltages with a seven-phase connexion. The voltage vectors form the sides of a seven-sided polygon, and the equivalent star voltage is given by the radius of the circumscribing circle.

It is clear that

$$E = \frac{E_{AB}}{2 \sin \pi/m},$$ (27)

where m is the number of phases.

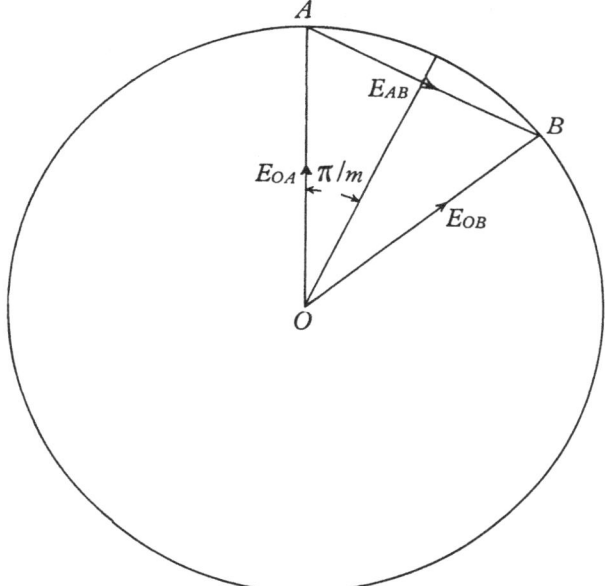

Fig. 29. Equivalent star voltage of seven-phase mesh

The frequency f of the voltage and current depends on the speed of rotation n_0 of the flux and is given by

$$f = n_0 p/2,$$ (28)

where p is the number of poles of the machine. This frequency is the same as that of the voltage induced by the same flux in a winding on the stator, and it therefore follows that a stator winding can be connected in the same circuit as the commutator winding. It is also the frequency of the current which has to flow in a stator winding in order to set up a flux rotating with speed n_0. Thus an A.C. commutator machine may be excited by means of a stator winding connected either in shunt or in series with the main commutator winding.

7. The current distribution and magnetomotive force

The wave of M.M.F. set up by a polyphase armature winding is related to the phase current flowing in at the brushes, the number of phases and the winding distribution. The relationships can be explained most clearly in terms of a three-phase winding. In such a winding, the currents flowing in at the brushes at any instant divide between the sections of the armature winding in the same way as in any three-phase delta-connected winding. Although the actual conductors which form the phases of the delta are continually changing, the distribution of the currents in space does not depend on the position of the armature, but depends solely on the currents in the brushes. For a balanced three-phase system of currents in the brushes, the magnitude and phase of the currents in internal sections of the winding are given by the vector diagram in Fig. 28(c), in which O is the centre point of an equilateral triangle ABC, the sides of which represent the line currents I_A, I_B and I_C. The currents I_{BC}, I_{CA} and I_{AB} are represented by the lines AO, BO and CO because the vector difference of the currents in two sections which join at a brush must equal the external brush current. In a similar way, a polygon can be drawn for any number of phases.

Distribution of current round the machine

In the previous section, it was explained that the three-phase armature winding differs from the ordinary induction motor winding because each layer of conductors is divided into three sections per pair of poles each extending over two-thirds of a pole pitch, instead of six sections each occupying a third of a pole pitch. The current distribution therefore differs somewhat from that of an induction motor winding, particularly when the pitch of the coils is shortened.

In Fig. 30, which represents the same full-pitch winding as Fig. 26, the current in each coil of the winding is indicated by arrows. The current I_{AB} is represented by one arrow, I_{BC} by two, and I_{CA} by three. The arrows show that the two conductors belonging to top and bottom layers respectively, which occupy the same slot, carry currents of different phases; the total current in the two conductors is therefore equal to one of the brush currents. For example, the conductors in line with the position of the brush A carry currents I_{AB} and I_{CA}, so that the total current in the pair of conductors lying in the same slot is equal to I_A, which is the vector difference of I_{AB} and I_{CA}. Consider the instant when the current in brush A is maximum; the resultant current in each pair of conductors is then indicated by the arrows

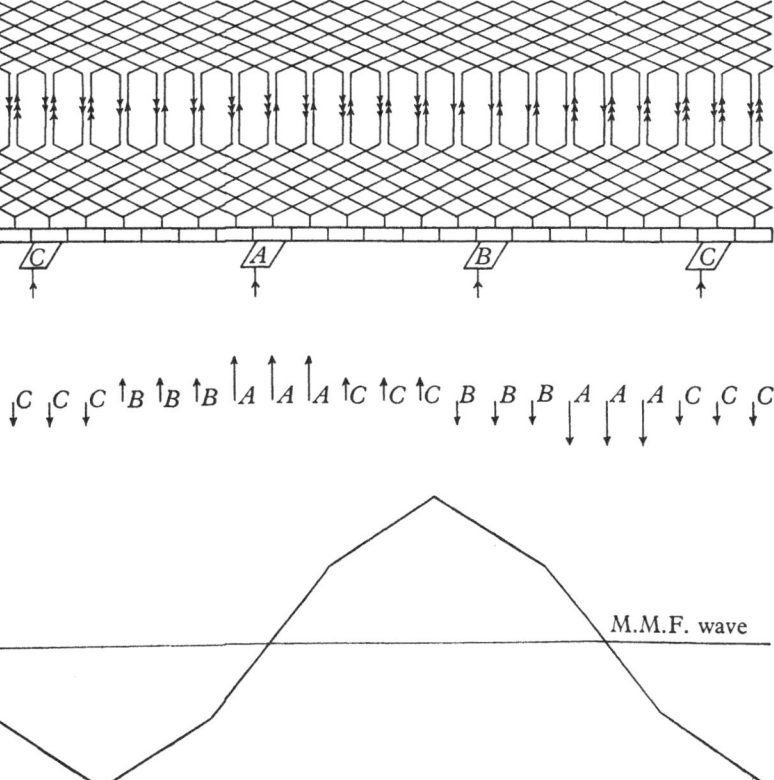

Fig. 30. Armature currents and armature M.M.F.

below the winding diagram, where the length and direction of each line represents the magnitude and direction of the current. The current distribution is therefore exactly equivalent to that in an ordinary three-phase winding with three sections per pole, each carrying the brush current, but having one-half of the actual number of conductors.

A full-pitch three-phase armature winding thus produces a rotating M.M.F. wave of approximately sinusoidal shape, as in a symmetrical induction motor winding. Although the actual conductors connected between a pair of brushes are continually changing as the armature rotates, the current distribution in space at any instant depends only on the instantaneous currents flowing in at the brushes, and hence the speed of rotation of the M.M.F. wave is the speed corresponding to the frequency of the currents. It can thus be seen that the M.M.F. wave, produced by three-phase currents of a given frequency, rotates at the same speed as a rotating flux which induces voltages of that frequency at the brushes. From the point of view of the frequency of external voltages and currents, and the speed of rotation of flux and M.M.F. waves, a three-phase commutator armature winding behaves in exactly the same way as a stationary three-phase winding on the stator.

When the coil pitch of a three-phase winding is shortened, the distribution of the current becomes unsymmetrical, that is to say, conductors at corresponding points on adjacent poles do not carry equal and opposite currents. This has the effect of distorting the M.M.F. wave by introducing harmonics which may be large enough to cause trouble if the coil pitch is much less than 100%. Unless the armature winding M.M.F. is small, or is neutralized at every point by means of a compensating winding, the coil pitch of a three-phase armature winding should not be shortened, except for the small amount necessary for using the 'stepped' type of winding (see p. 93).

When a commutator winding is not directly connected to a supply system, it is often advantageous to use more than three phases. (It should be remembered that the normal two-phase arrangement is really a four-phase connexion.) This may be done in order to obtain a convenient number of brush studs on the commutator, to reduce the reactance voltage of commutation, or to eliminate harmonics in short-pitch windings. When the number of phases is even, the current distribution is always symmetrical even when the pitch is shortened. When the number of phases exceeds three and is odd, the current distribution is unsymmetrical if the pitch is shortened, but the number of different orders of harmonics as well as their magnitudes are less than in a three-phase winding. However, whatever the number of phases and whatever the coil pitch may be, the frequency of the currents and the speed of rotation of the fundamental M.M.F. wave are related in the same way.

Calculation of magnetomotive force

The maximum value $AT_{\text{max.}}$ of the fundamental M.M.F. wave can be calculated in terms of the current I_c in the coils, and the total effective number of turns in series between the brushes T_a, in the same way as for an induction motor winding.

$$AT_{\text{max.}} = \frac{2\sqrt{2}}{\pi}\frac{m}{p}I_c T_a. \qquad (29a)$$

Or, if I_b is the brush current and T_e the effective number of turns in the equivalent star connexion as defined on p. 41 when considering the induced voltage, then

$$AT_{\text{max.}} = \frac{2\sqrt{2}}{\pi}\frac{m}{p}I_b T_e. \qquad (29b)$$

Armature acting alone

Consider, as an example, a machine where the armature winding is the only one present. Such a machine is known as a 'Leblanc exciter', and is one of the earliest forms of phase advancer. In this machine the rotating flux produced by the armature current induces a voltage in the winding which is 90° out of phase with the current, and lags or leads according to the relative direction of rotation between armature and flux.

The result may be deduced as follows. When the armature is stationary, it is obviously equivalent to a three-phase reactor. Rotation of the armature affects the magnitude of the voltage but not its frequency or its phase. Its phase is not affected because the position of the flux and therefore the phase of the voltage depends only on the phase of the current. Hence the voltage remains 90° out of phase with the current, but its magnitude and also its sign depend on the relative speed between armature and flux. If the armature rotates in the opposite direction to the flux, the voltage increases compared with its value at standstill and remains reactive. If the speed is raised from zero in the same direction as the flux, the voltage first decreases, and then becomes zero when the relative speed between armature and flux is zero.

If the speed is increased above this value, the voltage reverses and becomes capacitative.

When the armature runs in the same direction as the field and at a higher speed, the Leblanc exciter is thus equivalent to a capacitor and if it is connected in the rotor circuit of an induction motor, it can be used to advance the phase of the rotor current.

Leakage reactance of a commutator winding

The leakage reactance of a commutator winding depends partly on the frequency of the current in the brushes, and partly on the speed of rotation of the machine. The majority of the conductors at any instant carry currents which are changing at the external frequency, that is, the frequency of the brush currents. A few conductors, on the other hand, are short-circuited by the brushes, and the currents in these conductors undergo a more rapid rate of change, which depends on the speed of rotation of the armature.

With the armature at standstill the reactance is proportional to the external frequency and can be calculated in the same way as for a normal three-phase winding. This component of reactance remains unchanged when the armature rotates, and is equal to $K_a n_0$, where K_a is a constant. The reactance due to the short-circuited conductors, on the other hand, is proportional to n and can be expressed as $K_c n$, where K_c is another constant. Hence the total leakage reactance of the winding is

$$X = K_a n_0 + K_c n. \tag{30}$$

The magnitude of K_c relative to K_a depends on the conditions in the short-circuited coils, that is, on the way in which the currents change during commutation. The values to be used depend on the particular type of machine.

8. Compensating windings

In large D.C. machines it is often necessary to provide on the stationary field magnet a distributed compensating winding which neutralizes the armature M.M.F. at every point. This prevents the distortion of the flux wave which occurs on load in ordinary direct current machines. Certain types of A.C. commutator machine, particularly those which are excited by means of stator windings, also use compensating windings.

In a D.C. machine the armature M.M.F. wave has its maximum value at fixed points between the main poles, and its chief effect is to increase the flux density on one side of the pole and reduce it on the other. In a polyphase commutator machine on the other hand, the M.M.F. wave rotates and the position of maximum M.M.F. is constantly changing. As shown in the last section, the armature M.M.F., if acting alone, would produce a flux in such a phase position as to induce a voltage 90° out of phase with the armature current. Therefore whenever this action would interfere with the operation of the machine it is necessary to use a compensating winding.

Resultant induced voltage in armature and compensating windings

The magnitude of the voltage induced in the armature winding is proportional to the effective number of turns T_e of the winding and to the relative speed between the flux and the armature, in accordance with equation (25). It can therefore be considered to be made up of two components, one depending on the speed and one depending on the external frequency. The voltage induced in the compensating winding is exactly equal and opposite to the second component of the armature induced voltage, if the compensation is exact, that is, if the effective number of turns of the compensating winding is also T_e. The resultant voltage induced in the two windings in series is therefore equal to the first component, which is independent of the frequency and depends only on the flux and on the speed of the armature.

The result can be expressed as follows:

Armature induced voltage $= \pi\sqrt{2}\,\Phi T_e(n-n_0)\,(p/2)10^{-8}$, $\quad(31\,a)$

Compensating winding voltage $= \pi\sqrt{2}\,\Phi T_e(n_0)\,(p/2)10^{-8}$, $\qquad(31\,b)$

Total induced voltage $\qquad = \pi\sqrt{2}\,\Phi T_e n(p/2)10^{-8}$. $\qquad(31\,c)$

There are in addition comparatively small voltages induced by the leakage fluxes of both windings which depend on the frequency of the currents and on the currents themselves. As in other types of machine these can be taken into account by introducing a leakage reactance into the circuit.

A compensating winding can obviously only be used in a machine with fixed brush gear. It is always used in machines where it is required to make the induced voltage depend solely on the exciting current in a stator winding.

Errors in compensation

It is instructive to examine the effect of an error in compensation, as it is not usually possible in practice to make the M.M.F.s of the armature and compensating windings balance exactly. The error may be introduced in two ways:

(1) The effective numbers of turns may not be exactly equal. This error depends on the initial design and is difficult to change when the machine is made. Its effect is to set up a small resultant M.M.F. which produces a voltage in time quadrature with the current.

(2) The axes of the two windings may not coincide. This error depends on the brush position and can be readily adjusted by moving the brushes. The M.M.F. due to it is displaced by 90° in space from that produced by the armature winding alone and induces a voltage in phase with the current.

In Fig. 31, if OA represents the armature M.M.F., CO the compensating winding M.M.F., and the angle AOC the displacement of the brushes in electrical degrees forward in the direction of rotation from the *neutral position* in which the two M.M.F. waves are exactly opposite, then the resultant M.M.F. is CA. The flux due to the component NA in phase with OA induces a voltage in phase with that which would be produced by the armature acting alone, that is, the voltage is 90° out of phase with the current and lags behind the current if the armature runs faster than the field and in the same direction. Since, if we neglect saturation and assume the armature speed to be constant, the voltage is proportional to the current, this effect is equivalent to a reactance, which may be either inductive or capacitative, but which, unlike most reactances, is independent of the frequency.

Fig. 31. Vector diagram of armature and compensating M.M.F.'s

The flux due to the component CN induces a voltage which is 90° out of phase with that due to the component NA and therefore in phase with the current. If the brushes are moved forward from neutral in the direction of rotation the result is to increase the effective resistance of the machine, which then acts as a series motor and delivers mechanical power to the shaft. If, on the other hand, the brushes are moved backward from neutral, the effective resistance due to the series excitation of the armature and compensating windings becomes negative, and the machine acts as a series generator. This, however, leads to conditions of instability, and it is therefore necessary in this type of machine always to keep the brushes on the forward side of neutral.

Owing to the difference between the effective numbers of turns of the two windings and the difference in phase position, which introduce these errors, the result obtained on p. 49 is not strictly accurate. However, when the error in compensation is small, it is still approximately true to assume that the voltage due to an exciting current is independent of the frequency, and that the total induced voltage is obtained by superimposing on this main voltage a voltage proportional to and at a fixed phase angle to the current, to allow for the error in compensation.

Chapter 3

TYPES OF POLYPHASE COMMUTATOR MACHINE

9. General classification

Although many different arrangements of polyphase commutator machines are possible, the variations depend mainly on the external connexions rather than on the machine itself. The machine as such is a combination of two or more windings of the types discussed in the last chapter, and when considered in terms of their windings, the polyphase commutator machines can be classified into three main groups. The common features, which distinguish them from other types of electrical machine, are the commutator winding on the rotor, and the polyphase brush gear between which polyphase voltages are generated. The three groups differ between themselves in the arrangement of other windings on rotor and stator, and the way in which these are connected.

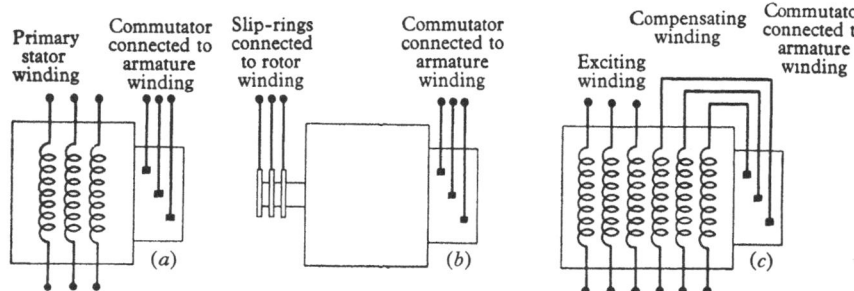

Fig. 32. Classification of polyphase commutator machines

Fig. 32 shows the basic connexion of each of the three groups. In order to give a consistent treatment, a uniform method of representation is adopted throughout this book. In each diagram of Fig. 32, the main rectangle represents the stator, and the three-phase windings indicated inside the rectangle are stator windings. A smaller rectangle to the right of the main rectangle represents the commutator with square dots to indicate the brush connexions.

Slip-rings are shown at the side of the stator rectangle in Fig. 32(b) and in later diagrams of induction motors. The commutators and slip-rings imply the presence of corresponding rotor windings which are not specifically indicated.

It has been shown that stator windings can be connected to the commutator brush gear because the frequency of the voltage and current corresponding to a given rotating field is the same in both. The frequency of the voltage between slip-rings is, however, different from that at the commutator brush gear, and these therefore cannot be connected in the same external circuit.

The essential features of the three types of machine represented in Fig. 32, are the following:

(a) *Polyphase shunt and series motors.* In addition to the commutator winding there is a stator winding similar to that of an induction motor. The machine resembles a polyphase induction motor except that the rotor winding is of the commutator type.

(b) *Commutator frequency changer.* The frequency changer has both a commutator and a set of slip-rings, sometimes connected to two separate rotor windings, and sometimes connected internally to the same winding, as in a rotary converter. In its simplest form, the frequency changer has no stator winding.

(c) *Scherbius machine.* The Scherbius machine is a polyphase A.C. generator in which the output voltage is generated in a rotor winding of the commutator type. The rotating field (or its equivalent) is set up by a polyphase exciting winding on the stator. In addition, a compensating winding, connected in series with the brushes so as to neutralize the armature reaction field, is provided on the stator. The arrangement is similar to that of a compensated D.C. generator except that there are three supply leads instead of two.

In each case, additional windings may be provided for special purposes, or windings may be combined together. For example, interpole or damping windings may be used to improve commutation. Again, a commutator frequency changer combined with a stator winding in a particular way becomes the Schrage motor, which is, in effect, an induction motor and a frequency changer combined into a single machine.

10. Polyphase shunt and series motors

The stator winding of a shunt motor or of a series motor is connected to the supply voltage and is considered to be the primary winding of the machine. The flux of the machine rotates at the synchronous speed corresponding to the frequency and the number of poles.

The commutator winding is considered to be the secondary of the machine. The magnitude of the voltage induced in the armature winding depends on the magnitude of the rotating flux, and on the relative speed between flux and conductors, that is, on the slip, just as in the induction motor. In fact, if the brushes are short-circuited, the machine then behaves like an induction motor with short-circuited slip-rings. Unlike the induction motor, however, the frequency obtained at the secondary terminals is always the same as the supply frequency, whatever the speed of the rotor may be. Although the voltages induced in individual conductors are at slip frequency, the voltage which appears at the brushes is converted to supply frequency by the action of the commutator. Because of this property, the secondary winding can be connected to the stator winding, either directly or through a transformer, and a much wider range of characteristics can be obtained than with an induction motor.

Apart from the question of frequency, the fundamental relationships between the voltages, currents, flux and slip are exactly the same as those given in Chapter 1 for the induction motor. In addition, however, there is a relationship between the stator and rotor voltages or currents, depending on how the stator and rotor are interconnected. Different results are obtained according to whether the effective connexion of the rotor is in shunt or in series with the stator.

Polyphase shunt motor

Typical connexions for a shunt motor are given in Fig. 33. The stator winding is connected directly to the supply, and for this reason the shunt motor is often called the 'stator-fed' shunt motor, to distinguish it from the 'rotor-fed' Schrage motor, which is sometimes called (rather inaccurately) a 'shunt motor'. The flux is approximately constant as in an induction motor. The rotor is connected through a transformer, or equivalent apparatus, to the same supply, and therefore, apart from impedance drops in the transformer, has a constant voltage impressed on it. Hence the shunt motor is equivalent to an induction motor with a constant injected voltage.

The magnitude and phase of the injected voltage determine the speed and power factor at which the motor will operate at any load. With a given setting of the transformer or regulator, the

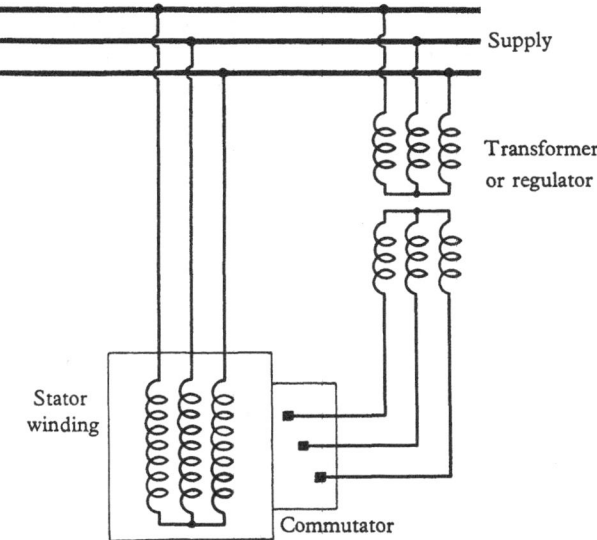

Fig. 33. Connexions of polyphase shunt motor

speed will not change greatly as the load varies; that is, the motor has shunt characteristics similar to those of a shunt-excited D.C. motor. To adjust the speed, the injected voltage must be varied, particularly the component in phase with the secondary induced voltage of the motor. If the transformer secondary voltage is varied between positive and negative values, the speed of the motor will vary between sub-synchronous and super-synchronous values.

Early examples of shunt motors were controlled by transformers with tap-changing switch-gear. It is, however, preferable to obtain a continuous control of speed instead of a limited number of steps. Moreover, the secondary current is generally high because the voltage is limited by consideration of commutation, and there are often more than three secondary phases. A transformer with tap-changing gear therefore becomes very cumbersome, and modern shunt motors are almost always controlled by means of an induction regulator, sometimes acting by itself, and sometimes in conjunction with a transformer, or, what is equivalent, an auxiliary winding on the motor stator, or on the regulator stator.

The brushes of the shunt motor are generally held fixed, but sometimes brush movement is introduced in order to control the relative phase of the injected voltage. Details of the arrangement of the controlling apparatus are given in Section 24.

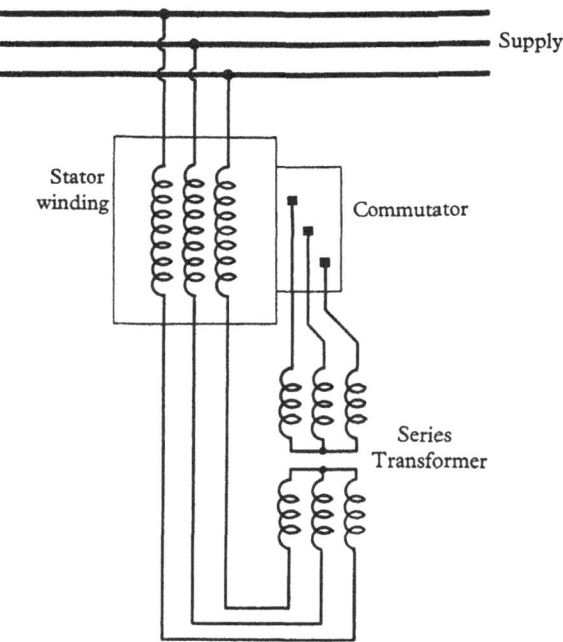

Fig. 34. Connexions of polyphase series motor

Polyphase series motor

Fig. 34 shows the connexions of a polyphase series motor. Theoretically it would be possible to make a direct series connexion between the stator and rotor, but a transformer is generally necessary in practice because of the low voltage at the commutator. The transformer normally has a fixed ratio, and control of the motor speed is obtained by moving the brush gear.

In the series motor, with the brushes in a given position, the flux, and hence the torque, depends only on the current flowing in the stator and rotor windings in series. The current varies considerably as the speed changes, being much greater at low speeds when the rotor voltage is greatest. Hence the torque-speed characteristic of the polyphase series motor is similar to that of a D.C. series motor, and shows a large torque at low speeds and a small torque at high speeds.

With a different brush position a different torque-speed characteristic is obtained. Thus speed variation is obtained by movement of the brush gear, which gives continuous control without introducing any loss of power like the loss in the resistances used for controlling a D.C. series motor.

11. The commutator frequency changer and the Schrage motor

Frequency changer

The slip-rings of a commutator frequency changer are generally connected to a supply of constant voltage and frequency, and the machine is used as a source of voltage at a different frequency. The output voltage, which is approximately constant, is taken from the commutator brush gear, and has a frequency which depends on the speed of rotation of the rotor as well as the frequency of the input voltage.

The flux in the machine rotates relative to the rotor at the synchronous speed corresponding to the supply frequency and the number of poles. The actual speed of rotation of the flux is the difference between the synchronous speed and the speed of rotation of the rotor. The frequency of the output voltage at the commutator brush gear, depends on the speed of the flux, and is therefore the same as the slip frequency of an induction motor which has the same number of poles, runs at the same speed, and is supplied at the same frequency, as the frequency changer. Hence if the frequency changer is mechanically coupled to an induction motor having the same number of poles, it becomes a source of slip-frequency injected voltage which can be used for controlling the speed or power factor of the induction motor.

The phase sequence of the connexions between the supply and the frequency changer slip-rings must be such that the flux in the frequency changer rotates backwards relative to the rotor, when the induction motor is running in a forward direction. The frequency of the output voltage of the frequency changer is then always equal to the slip frequency of the motor irrespective of the speed of rotation. The secondary connexions between the frequency changer and the induction motor slip-rings must also be made with correct phase sequence.

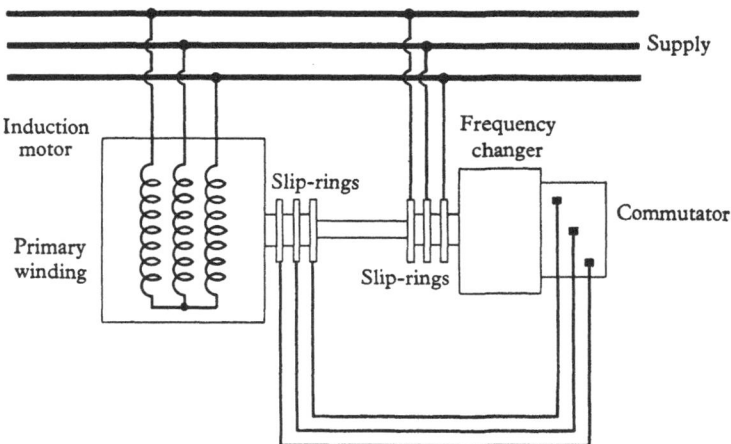

Fig. 35. Connexions of induction motor with commutator frequency changer

Fig. 35 shows the connexions of an induction motor with a commutator frequency changer. Apart from the impedance drops in the frequency changer, the induction motor has a constant injected voltage, and therefore operates with shunt characteristics, at a speed which depends on the in-phase component E_{kp} of the injected voltage. To adjust the speed, the voltage obtained from the frequency changer must be varied either by supplying the input voltage from a variable ratio transformer or regulator, or by providing two sets of movable brushes on the commutator as explained later.

The commutator frequency changer, which is a large and expensive machine for its output, is not often used directly as a regulating machine for an induction motor. It is much more often used as an exciter for a Scherbius machine, where the same requirement of providing a slip-frequency voltage applies, but where the output required is much less. The most important application of the principle of operation of the frequency changer is the Schrage motor.

The Schrage motor

The connexions of the Schrage motor are given in Fig. 36. The slip-rings are connected to the polyphase supply, and the commutator is provided with two sets of brushes mounted on separate rings each of which can be moved round the commutator. In addition there is a stator winding of which one end of each phase is connected to one set of brushes, while the other ends are connected to the other set of brushes.

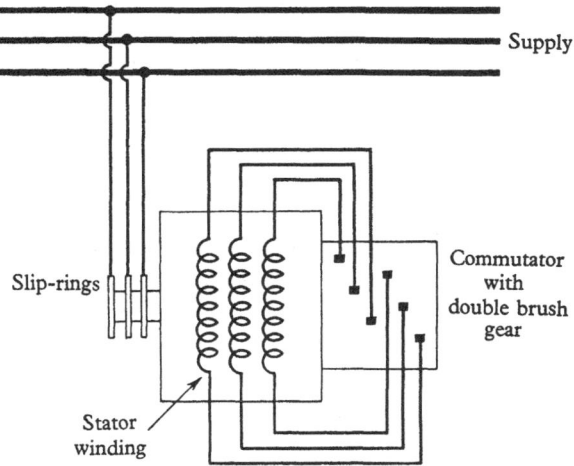

Fig. 36. Connexions of Schrage motor

The slip-ring winding and the stator winding together constitute an induction motor in which the normal arrangement of stator and rotor windings is inverted. The rotor winding is the primary of this induction motor and the stator winding is the secondary. The slip-ring winding and the commutator winding together constitute a frequency changer, in which the flux is the same as that of the induction motor. Thus the frequency of the voltage obtained at the commutator brush gear is the slip frequency of the induction motor. In the Schrage motor the commutator winding provides the injected voltage which is applied to the secondary winding on the stator.

With any given setting of the brush gear the injected voltage is approximately constant, and causes the motor to have a shunt characteristic. By moving the two sets of brush gear, the magnitude and phase of the injected voltage can be adjusted, thus controlling the speed and power factor of the motor. The Schrage motor is approximately equivalent to an induction motor with a frequency changer.

12. The Scherbius machine

The Scherbius machine can best be understood by comparing it with a D.C. generator. It generates polyphase voltages, generally at a low frequency, between the commutator brushes in the same way that a D.C. machine generates D.C. voltages. The flux which induces the armature voltage is set up by the field current in an exciting winding provided on the stator, and the output voltage, obtained when the machine is running at a given speed, is determined by the exciting current. There is a close analogy between the Scherbius machine and the ordinary D.C. generator.

Uniformly slotted machine

The Scherbius machine is generally constructed with salient poles and interpoles—like a D.C. machine except that the stator magnetic circuit is laminated throughout. The term 'Scherbius machine' is often understood to refer to this arrangement only; but, for small powers, machines are frequently built, which, although they have uniformly slotted punchings, still come within the definition of a Scherbius machine given in Section 9; the series and shunt-excited phase advancers described in Chapter 9 are often built in this way. In a uniformly slotted machine of the Scherbius type, both the exciting and compensating windings are normal polyphase windings, and the flux is a rotating flux whose speed corresponds to the frequency of the exciting currents. As the uniformly slotted machine is simpler and more fundamental, it will be considered first.

The magnitude, frequency and time phase of the exciting current determine the magnitude, speed of rotation and space phase of the rotating flux wave, and hence the magnitude, frequency and time phase of the output voltage on open circuit. The frequency of the output voltage is necessarily always the same as that of the exciting current. Its magnitude is related to that of the exciting current by the magnetization curve of the machine. There is a constant phase angle between the exciting current and the output voltage, depending on the relative position in space of the exciting winding and the brush gear. Hence the output voltage can be controlled by controlling the exciting current, just as in a D.C. generator.

A D.C. generator can operate without a compensating winding, but in the Scherbius machine a compensating winding is essential, for two reasons. First, the armature reaction field due to load currents would induce a voltage 90° out of phase with the current, and would interfere with the operation of the machine. Secondly, it is essential to keep the exciting ampere-turns as small as possible, because, even at low frequencies, the reactive drop in the exciting winding may be many times the resistance drop with a consequent increase in the exciting voltage. Hence, to avoid having to supply an excessive kVA. input to the exciting winding, the air-gap must be a good deal shorter than that of a D.C. machine.

Salient pole machine

In a uniformly slotted Scherbius machine the rotating flux cuts the conductors undergoing commutation, and the possible output is severely limited. For larger outputs the salient pole construction shown in Fig. 37 is used. The stator laminations are made by notching out large and small slots as shown, so as to form alternate main poles and interpoles. The three-phase machine represented in the diagram has six salient poles marked *A*, *B*, *C*, and is equivalent to a uniformly slotted machine with four poles, that is, with two complete flux waves. Each pole pitch thus covers 120 electrical degrees, instead of the 180 electrical degrees covered by the poles of a D.C. machine, or of the single-phase A.C. commutator motors used for traction work. In the Scherbius machine, in order to bring the two sides of a coil undergoing commutation under the interpoles, the pitch of the armature coils must be approximately 120 electrical degrees, or $66\frac{2}{3}\%$ of full pitch. The exciting and interpole windings consist of coils wound round the main poles and interpoles respectively, as in a D.C. machine, and the compensating winding is distributed over the pole face.

Other possible arrangements have been suggested and tried, but that shown in Fig. 37 is the most commonly used and generally the most satisfactory. The reasons for the arrangement are discussed in Section 15. Many interesting points arise in connexion

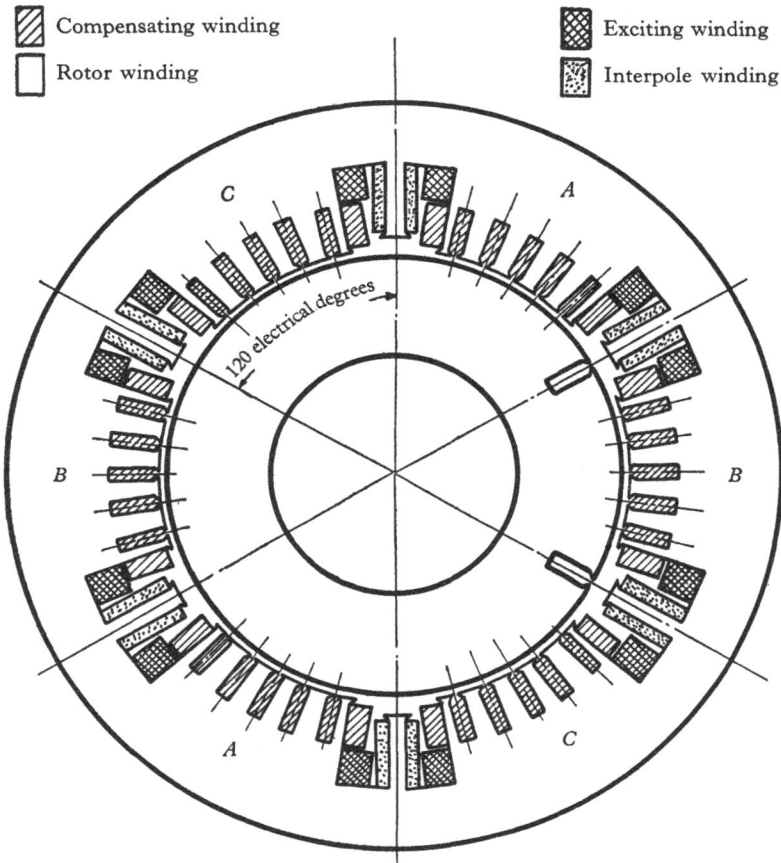

Fig. 37. Windings in a Scherbius machine

with the details of the windings which cannot be gone into here. When considering the operation of the Scherbius machine, however, particularly in relation to its external connexions and the control schemes with which it is used, it is best to think in terms of an equivalent uniformly slotted machine with a rotating field.

Although the flux distribution in a salient pole machine is a good deal more complicated, the fluxes in the three poles corresponding to the three phases do rise and fall in sequence, and the result is equivalent to that obtained with a rotating field. In the diagrams explaining the uses to which the Scherbius machine is put, the interpole winding is not shown.

External connexions

The Scherbius machine can be used as an independent source of low frequency power, or as an exciter for another Scherbius

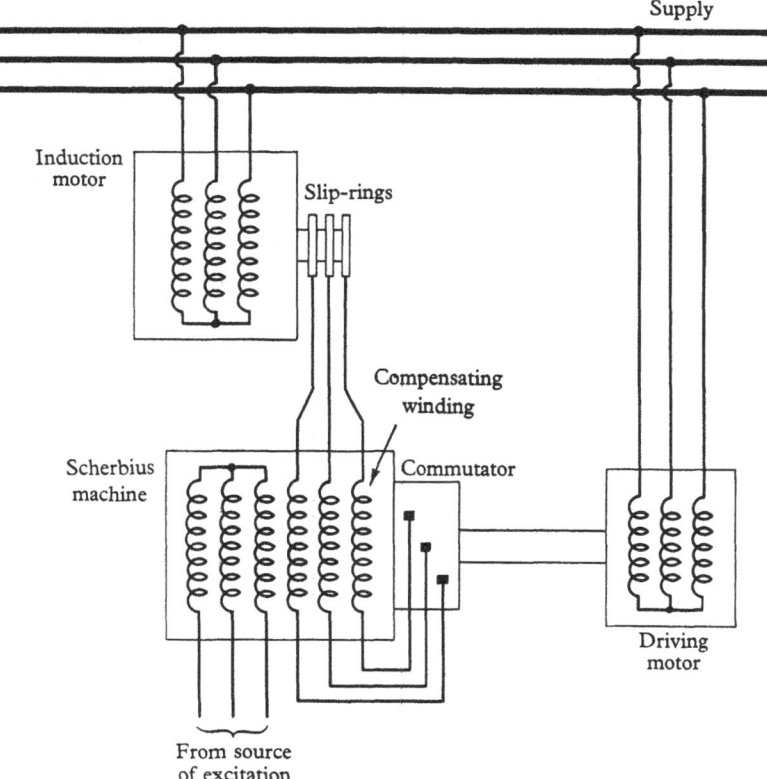

Fig. 38. Connexions of induction motor controlled by Scherbius machine

machine, but its commonest use is as a source of the injected voltage required for controlling the speed or power factor of a slip-ring induction motor, as indicated in Fig. 38. The main terminals coming from the armature and compensating winding in series are connected to the slip-rings of the main induction motor. The exciting winding is supplied from a source of excitation which is not shown. The frequency of the exciting current must always agree exactly with the slip frequency of the induction motor, but, subject to this condition, complete control of the induction motor is obtained by controlling the exciting current of the Scherbius machine. The practical methods of providing and controlling this exciting current are discussed in Chapter 8.

The Scherbius machine may be driven in any convenient manner, as the frequency of the generated voltage is quite independent of the speed of rotation. In Fig. 38, it is driven separately by an A.C. motor, which may be a slip-ring or squirrel-cage induction motor, a synchronous motor, or a commutator motor. It can also, if convenient, be directly coupled to the main induction motor. The direction of flow of power in the Scherbius machine and its driving motor depends on whether the main machine being controlled by it is motoring or generating, and on whether the speed of the main machine is below or above synchronism, as explained in Chapter 1. For sub-synchronous operation as a motor, for example, the Scherbius machine takes power from the slip-rings and acts as a motor, while the machine coupled to it acts as a generator and returns power to the supply.

Chapter 4

COMMUTATION IN THE A.C. COMMUTATOR MACHINE

13. The general problem of commutation

One of the principal problems in designing any machine in which current is collected from a commutator, is that of obtaining good commutation. The final test of good commutation is the amount of wear of commutator and brushes, which affects the length of life of the machine and the cost of maintaining it in good running condition. Destructive commutation is generally accompanied by visible sparking at the brushes, and it is usual to test the quality of the commutation of a machine by examining the sparking produced under running conditions. The dissipation of energy at the brush contact is also closely related to the quality of the commutation.

In a d.c. machine the main problem to be dealt with in connexion with commutation arises from the fact that the current in a coil has to reverse when the segments to which it is connected pass under a brush. The M.M.F. due to the current sets up a magnetic flux round the slot in which the conductor is situated, and a sudden change in the current causes a sudden change in the flux embracing the conductor, so that a voltage which opposes the change in current is induced in the coil. Unless, therefore, a voltage of sufficient value in the opposite direction is present in the coil undergoing commutation, the current cannot change to its final value before the coil is open-circuited at the brush contact, and sparking results. The voltage which would be induced between adjacent segments, if the current changed uniformly during the period while the coil in question is short-circuited by the brush, is known as the *reactance voltage of commutation*. The phenomenon in general is referred to as *current commutation*.

In an A.C. commutator machine, a sudden change of current has still to be effected from its value in one phase to its simultaneous value in the next phase in the armature winding as the segments connected to a coil pass under a brush, and the problem of current commutation is similar to that in a D.C. machine. In addition, there is the further difficulty in an A.C. machine that a voltage— known as the *transformer voltage of commutation*—may be induced by the main alternating or rotating flux of the machine in the coils short-circuited by the brushes.

The winding shown in Fig. 26 is reproduced in Fig. 39 with the important difference that each brush is bridging two segments instead of being in contact with only one. Thus the coils represented in the diagram by heavier lines are each short-circuited by a brush, and it can be seen that for a full-pitch three-phase winding the coil sides of the short-circuited coils are equally spaced a third of a pole pitch apart. The transformer voltage is

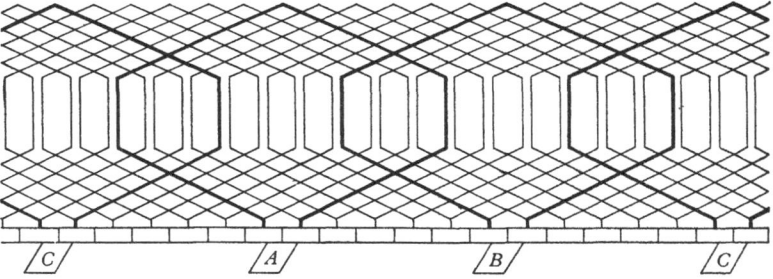

Fig. 39. Coils short-circuited by brushes

induced in these coils because they are linked with the main alternating or rotating flux. If the brush makes contact with more than two segments at any instant, more than one coil is short-circuited, and the voltage set up between the tips of the brush is correspondingly greater.

This effect, referred to as *voltage commutation*, does not arise in a normal D.C. machine because the main flux does not alternate, and because the brushes are so placed that the conductors making up the short-circuited coils are in a position between the poles where the main flux density is zero. Hence the total flux linking the short-circuited coils in a D.C. machine is not changing, and no transformer voltage is induced.

The most important factor in producing good commutation is the voltage drop set up between brush and commutator when current flows between them. The brush contact drop absorbs the reactance voltage and transformer voltage provided they are not excessive, and in many A.C. commutator machines the commutation depends entirely on the action of the brush. In other types of machine, these voltages are partly neutralized by means of interpole windings on the stator, and only a fraction of the voltage has to be absorbed by the brush contact. Commutation may also be assisted by the use of high resistance connectors or damping windings of various kinds to be described later.

14. Reactance voltage and current commutation

In a D.C. machine, the reactance voltage at a brush depends on the change of current in a coil short-circuited by the brush. The current I flowing in the brush divides equally between the two paths which join at the brush, so that the current in a coil changes from the value $I/2$ in one direction to $I/2$ in the other direction during the period of commutation. Hence, the total change in the current in the coil is equal to I, the current in the brush. If the brush current varies, the mean value of reactance voltage varies in proportion to the current.

It should be made clear that although the reactance voltage at any instant is being induced in a coil, the value of reactance voltage is related to the brush and not to any particular coil. Over a period of time, if the current flowing in the brush is constant, the reactance voltage at that brush is constant, although it is being continually transferred from one coil to the next. The constancy of the reactance voltage is based on the assumption, which is not strictly true, that the current in a coil changes uniformly during the period of commutation. In practice, the reactance voltage in a D.C. machine at a brush carrying a constant current, pulsates at a high frequency, and the calculated or measured value of reactance voltage is the mean of this pulsating value. The predominant frequency of pulsation is the frequency at which the segments pass under the brush, as a sudden change of current occurs every time a segment breaks contact with the brush. A periodic variation also occurs at slot frequency.

In an A.C. commutator winding, however, the brush current does not divide equally between the two paths which meet at the brush. At any instant the current divides into two unequal parts i_1 and

i_2, of which the algebraic sum $i_1 + i_2 = i$, where i is the instantaneous value of the current. By following round a coil in the same direction before and after the corresponding segments pass under a brush, it can be seen that the current changes from i_1 to $-i_2$. Hence the total change in current is $i_1 + i_2$ or i, the brush current.

Since the time of commutation during which the reactance voltage in a particular coil is induced is always short compared with the period of the A.C. cycle, the instantaneous value of the reactance voltage is proportional to the instantaneous current in the brush, as in the D.C. machine. Thus, in an A.C. commutator machine the reactance voltage per coil under a brush varies approximately sinusoidally with time and is proportional to, and in phase with, the current in the brush. In practice the curve of reactance voltage has a high frequency ripple superimposed on the fundamental voltage curve, as in the D.C. machine already mentioned. (See Fig. 55.)

Calculation of reactance voltage

Consider, as an example, a full-pitch three-phase lap-wound armature winding and brush gear, as illustrated by the developed diagram in Fig. 40. The diagram refers to the same type of winding as that shown in Fig. 39, but Fig. 39 does not show the arrangement of the conductors in the slots. In Fig. 40 each slot has six conductors, three in each layer, but, for the sake of clearness, only the coils in the two slots S_1 and S_2 are shown. The reactance voltage may be calculated approximately by the method given below, which can be used with simple modifications for any other winding pitch or number of phases, or for a wave winding.

Let i be the instantaneous current flowing in the brush B,

     q the total number of effective conductors per slot,

     w the number of turns per segment,

     P the permeance of the slot, that is, the ratio of the flux round the slot to the M.M.F.,

     t the slot pitch referred to the commutator diameter,

     c the segment pitch,

     b the brush thickness,

     d the commutator diameter,

     n the speed in revolutions per second.

In the winding shown in Fig. 40, where $q = 6$ and $w = 1$, only half the conductors in the slots S_1 and S_2 are affected by the commutation at the instant considered, because the brushes and coils are so spaced that the coil sides short-circuited by adjacent brushes do not lie together as they do in a D.C. machine. During the period over which a change of current is taking place in the slot, the current in half the conductors changes by the amount i.

The total change in ampere-conductors per slot is, therefore, $\dfrac{qi}{2}$, and the change in the flux is $\dfrac{4\pi}{10}\dfrac{Pqi}{2}$.

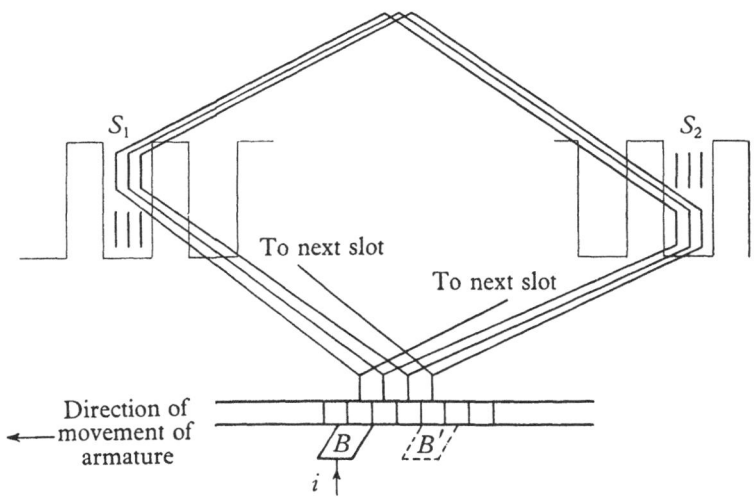

Fig. 40. Coils undergoing commutation

When the brush is in the position B relative to the commutator —as shown in full lines—the current in all the coils shown is flowing away from the brush, and none of the coils in the slots S_1 and S_2 is short-circuited by the brush. As soon as the commutator moves to the left from this position, one of the coils is short-circuited and the current begins to change. The change is complete when the brush has moved relative to the commutator to the position B' shown in broken lines. If the width of the insulation between the segments is neglected, the distance between the positions B and B' is $t - c + b$.

The time during which the change in flux takes place is therefore $\dfrac{t-c+b}{\pi dn}$. Hence the mean voltage between adjacent segments induced in w turns, or $2w$ conductors, is

$$e_c = 2w\frac{4\pi}{10}\frac{Pqi}{2}\frac{\pi dn}{t-c+b} \times 10^{-8}$$

$$= \frac{4\pi^2}{10^9}\frac{wqdnP}{t-c+b}i.$$

This expression, which is proportional to i, is the instantaneous value of the reactance voltage at the brush when the instantaneous value of current is i. If the current alternates according to a sine wave, of which the virtual value is I, then the reactance voltage per segment also alternates sinusoidally and its virtual value is

$$E_c = \frac{4\pi^2}{10^9}\frac{wqdnP}{t-c+b}I. \qquad (32)$$

The value of the permeance P depends on the shape of the slot and adjacent air-gap, on the length of the armature core, and on the length and position of the end windings. Its value can be calculated in the same way as for D.C. machines.

If the number of phases is greater than three, or if the pitch of the coils is shortened, the conditions are similar to the three-phase full-pitch case, unless the winding is such that conductors short-circuited by different brushes lie in the same slot. In such windings, of which the usual short-pitch armature winding of a salient pole Scherbius machine is an example, the change in M.M.F. in the slot is the sum of the changes of two M.M.F.'s of different phase, and the reactance voltage is increased. Another common example is a full-pitch winding with an even number of phases. Under symmetrical conditions, however, the total reactance voltage in the complete coil is still in phase with the brush current. Provided the condition where the coils overlap is avoided, the effect of increasing the number of phases is to reduce the reactance voltage at each brush, because the current per brush is reduced as the number of phases is increased.

It is evident that the rate of change of flux linking the slot, as the segments pass under the brush during the period of commutation, is not by any means uniform. The virtual value E_c of the reactance voltage merely gives a measure of the fundamental sine wave and does not indicate the high frequency pulsations which are superimposed on it. Nevertheless, while this value of reactance voltage is not the only criterion for current commutation, it is a very useful guide for purposes of comparison, provided that the general proportions of the machine do not differ widely from those normally used.

15. Transformer voltage and voltage commutation

The transformer voltage in the coil short-circuited by a brush is induced in the same way as the voltage induced in any other coil, by the alternation of the flux linking it or by its movement in the flux. In a machine with a uniform air-gap, in which a sinusoidal flux wave rotates at a uniform speed, the virtual value of the transformer voltage induced between a pair of adjacent segments—generally called the *transformer voltage per segment*—is

$$E_t = \pi\sqrt{2}K_p\Phi fw \times 10^{-8}, \tag{33}$$

where
$\Phi = $ flux per pole,

$f = $ frequency of induced voltage,

$w = $ number of turns per segment,

$K_p = $ the pitch factor of the coil.

The uniform air-gap, similar to that of an induction motor, is used in the majority of A.C. commutator machines on account of its simplicity. In motors which are regulated by shifting the brushes, no other arrangement is possible, and in such motors, no interpole winding can be used because the position of the short-circuited coil changes as the brushes move. Even when the brushes are fixed, interpoles often cannot be used. For example, in shunt and series motors the distortion of the main working flux would be too great. Both reactance voltage and transformer voltage must, therefore, be limited to a value which can be dealt with by the brush contact.

The transformer voltage causes circulating currents to flow in the short-circuited coils. The currents cause heating at the brush surface and sparking at the tip of the brush. These considerations set a limit to the permissible value of the transformer voltage.

Machines with uniform air-gaps and no interpoles normally use brushes of a thickness less than twice the pitch of a commutator segment; in such machines, the transformer voltage must not exceed about 2V. if the coils are connected directly to the commutator, or about 2·6V. if high resistance connectors are used. Now if the armature is moving relative to a constant rotating flux, the virtual value of the voltage induced in all coils is the same, whether they are short-circuited by the brushes or not. The voltage set up between brush studs of adjacent phases, which is the vector sum of the voltages in all the coils connected between the brush studs, is therefore limited to a comparatively low value, and for this reason, it is usual in this type of machine to have as many segments as is permissible from a constructional point of view. Compared with a D.C. machine, an A.C. commutator motor in which no interpoles are provided has a large number of thin brushes. The commutator voltage between points a pole pitch apart is usually not more than about 60 V.

Machines with salient poles

The transformer voltage in a machine with a purely rotating flux is proportional to the relative speed between flux and armature, that is, to the sum or difference of the speed of the flux and the speed of the armature. The total transformer voltage may therefore be split up into two parts, one a rotation voltage due to the movement of the conductors in the alternating flux at the points where the conductors lie, and the other a true transformer voltage due to the alternation of the flux linking the short-circuited coils. Since, in the machine with a uniform air-gap, the transformer action takes place through the medium of a rotating field, there is no object in considering the two parts separately. But if the stator core, instead of being uniformly slotted, is divided up into salient poles and interpolar spaces so that the flux is confined to the main salient poles, the rotation voltage in the short-circuited conductors due to the main flux is reduced to zero if they are located opposite the interpolar spaces, because the main flux is zero at these points. The true transformer voltage remains, however, owing to the alternation of the flux. When the effective speed of rotation of the flux is much less than the speed of the armature, as it usually is in machines used for low frequencies, the total transformer voltage is therefore much less than in a machine with a uniform air-gap. In other words, the voltage induced in conductors which lie under the main poles is much greater than the voltage induced in the short-circuited conductors, so that much higher terminal voltages can be generated when the salient pole construction is used. With this arrangement, it is not permissible to move the brushes; they must be so fixed that the short-circuited conductors lie opposite the interpolar spaces.

The necessary arrangement of the exciting winding for a full-pitch three-phase armature is shown in Fig. 41, which is a developed diagram representing 360 electrical degrees. The positions of the short-circuited conductors are indicated by the dotted lines 1, 2, 3, 4, 5 and 6, which, for a full-pitch armature winding, are equally spaced round the circumference. Opposite these points the stator is cut away to form six neutral zones in which the exciting coils lie. The six salient poles have round them exciting coils of successive phases as indicated by the letters A, B and C, distributed in the same order as the phase groups in a normal three-phase induction motor winding. From the point of view of the main voltage induced between the brushes on the commutator, the flux set up by this exciting winding is equivalent to a rotating flux, but it is evident that the M.M.F., and hence the flux density, is zero in each of the neutral zones. The change of M.M.F. from one neutral zone to the next is zero, and thus by symmetry the M.M.F. is zero at each of these points. Hence, as explained in the previous paragraph, no voltage is induced in the short-circuited coils due to the rotation of the armature in the flux.

The arrangement of salient poles and interpolar spaces is similar to that in a D.C. machine except that there are six poles instead of two in the space covered by a complete wave of flux, that is, in 360 electrical degrees. This is referred to as a *set of poles*. In the D.C. machine, there is no transformer voltage induced in the short-circuited coils because the flux in the poles does not alternate, but in the salient pole A.C. commutator machine, on the other hand, there is a transformer voltage induced by transformer action by the main flux. This transformer voltage is proportional to the frequency of the exciting current and is relatively small if the machine operates at a low frequency. The full advantage of using salient poles is only obtained when the whole of the flux in the machine is produced by exciting coils wound round the poles. It is, therefore, necessary for this reason, as well as for those already mentioned, to have a compensating winding on the stator which neutralizes the armature M.M.F. as nearly as possible at every point.

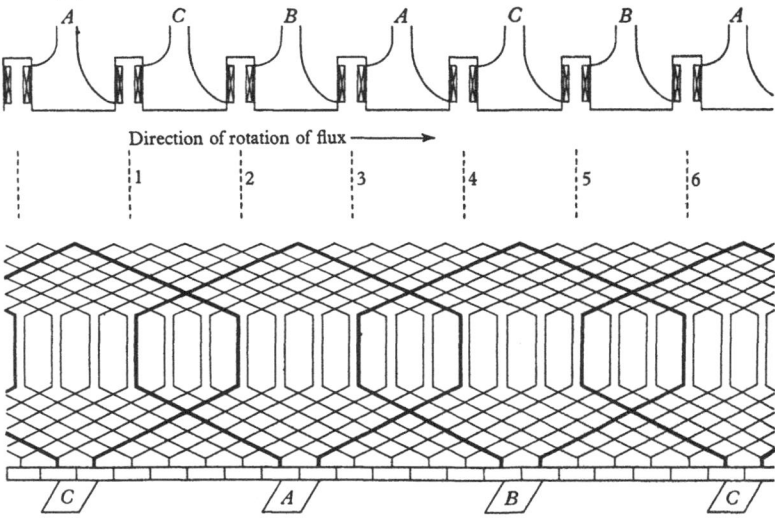

Fig. 41. Short-circuited coils in relation to neutral zones, 100 % pitch winding

With the six-pole arrangement, the size of the salient poles and the span of the exciting coils are very small, and it is more usual to construct this type of machine with three salient poles for each complete flux wave as shown in Fig. 42. The diagram, which shows a developed view of one set of poles with the corresponding armature winding and brush gear, represents the arrangement of windings usually employed in Scherbius machines, and agrees with that illustrated in Fig. 37. In order that the shorted conductors may come between the poles, the pitch of the armature winding is shortened to $66\frac{2}{3}\%$, and hence the opposite sides of the coils short-circuited by adjacent brushes fall together, and occupy only three points per set of poles as indicated by the dotted lines. With this arrangement, the dimensions of the poles and interpolar spaces are comparable with those usual in D.C. machines, and there is plenty of room for interpoles and interpole windings, while, at the same time, excessive leakage of flux between pole-tips is avoided.

In other types of A.C. commutator machines with fixed brush gear, in which the field is set up by a distributed exciting winding, the commutating conditions can be somewhat improved by cutting out wide grooves in the stator core, thereby increasing the air-gap

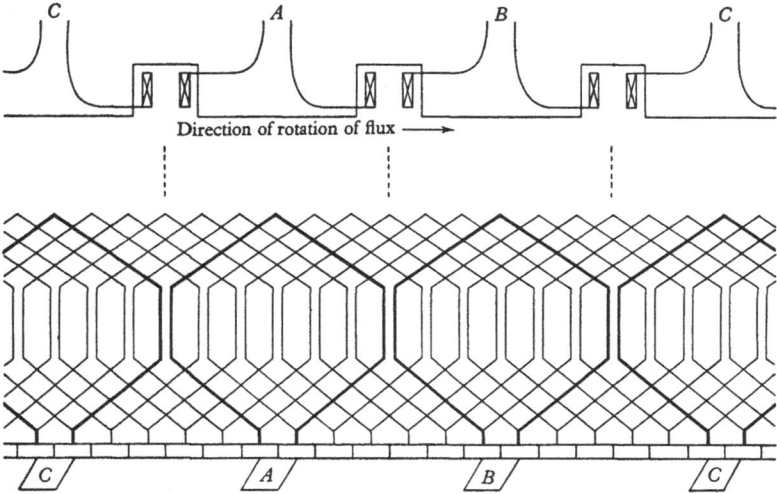

Fig. 42. Short-circuited coils in relation to neutral zones, 66⅔% pitch winding

at points opposite to the short-circuited conductors on the armature. The M.M.F. due to the exciting winding is still the same at these points as if there were no grooves, but the flux density is reduced owing to the reduced permeance across the air-gap. The voltage induced in the short-circuited coils by its rotation in the flux is, therefore, less than if the air-gap were uniform, but experience shows that the resultant effective transformer voltage is not reduced to the same extent as in a salient pole machine. The method is not very satisfactory in shunt and series motors, because of the distortion of the main flux; but it has been used successfully in commutator frequency changers both with and without interpole windings.

16. Interpoles

Neutralization of reactance voltage

Practically all D.C. machines are provided with interpoles situated between the main poles. Coils carrying the main armature current are wound round the interpoles and set up a local flux which induces a voltage in the coil short-circuited by the brushes, and neutralizes more or less completely the reactance voltage of commutation. The interpole flux must be opposite in direction to, and proportional to, the flux which would be produced by the armature winding acting alone at the point occupied by the interpole. Hence in an uncompensated D.C. machine the interpole ampere-turns are greater—usually about 25%—than the armature ampere-turns, in order to neutralize the armature M.M.F. at the point and set up a local flux in the opposite direction. When a D.C. machine has a distributed compensating winding which neutralizes the armature reaction, the interpole has only to carry the small number of turns necessary to set up the interpole flux.

In A.C. commutator machines in which the brushes are permanently fixed in one position, the reactance voltage can be neutralized by means of interpoles in exactly the same way. At every point occupied by a conductor undergoing commutation by a brush, an interpole can be provided with a winding carrying a current whose value at any instant is proportional to the instantaneous brush current and whose direction is such that the flux is opposed to that which would be set up at that point by the armature winding. When the brush current alternates, the interpole current must alternate also, and good correction is obtained over a wide range of loads and power factors if each interpole coil is connected in series with the corresponding brush.

This applies particularly to windings such as that illustrated in Fig. 41, in which each interpole is used to neutralize the reactance voltage at brushes of one phase only. When conductors which are short-circuited by brushes of different phases lie at the same point, as in the winding shown in Fig. 42, the interpole flux has to be out of phase with the current in any one brush in order to induce suitable voltages in the two conductors concerned which belong to different coils short-circuited by different brushes. In such a case, an interpole winding consists of two coils connected in different phases.

Practical details

With the construction normally employed for D.C. machines the provision of interpoles bolted to the magnet frame between the main salient poles is a simple matter, and the commutating flux can be readily adjusted by inserting shims of sheet iron between the interpole and the frame. In an A.C. machine on the other hand, the stator magnetic circuit must be laminated throughout, and interpole windings are wound in large slots notched in the punchings on either side of a tooth which forms the interpole. It is possible to adjust the commutating flux after the machine is built, by machining the face of each interpole, or by providing taps on the heavy interpole coils for changing the number of turns. These are cumbersome methods and, in practice, the machines have to be designed so as to give good commutation without any subsequent adjustment. At the same time, the use of three or more commutating zones in each flux wave instead of two makes the conditions somewhat easier than in a D.C. machine, because a double reversal of the current takes place in three or more stages instead of two.

In certain types of commutator machine which have no main stator winding—for example, the Leblanc exciter already mentioned—an interpole winding, which is then the only winding on the stator, may be used. To be fully effective the ampere-turns on each interpole must be greater than the armature ampere-turns at the point. In the majority of machines with interpoles, however, there is also a compensating winding, and the number of additional turns required to set up the local commutating flux is small. Generally the two windings are combined together in a single drum type winding, with the connexions arranged so that a few conductors are disposed on either side of the interpole, and are equivalent to a coil round the interpole.

When the main flux of the machine is small, as in most phase advancers, the reactance voltage may be neutralized in a simple manner by providing more turns in the compensating winding than is necessary for neutralizing the armature reaction. With this arrangement, in addition to the required local commutating flux, there is also a resultant main flux which produces a voltage between brushes proportional to the current. With the simplest possible construction, the stator is uniformly slotted and the compensating and exciting windings are both distributed windings of the ordinary type.

Neutralization of the transformer voltage

In addition to neutralizing the reactance voltage, the interpole flux may also be used to neutralize the transformer voltage in the short-circuited coils. For this purpose it is usual to provide additional windings on the same interpoles. For any condition of operation the magnitude and phase of the ampere-turns on each interpole can be chosen so as to set up the necessary local flux for inducing a voltage in the short-circuited coils equal and opposite to the transformer voltage.

With any given armature speed, the mean reactance voltage at any instant is always proportional to the instantaneous brush current, so that, apart from the high frequency pulsations and the error due to saturation of the interpole, complete compensation under all conditions of operation can be obtained by using a series interpole winding. The transformer voltage, on the other hand, generally depends on the frequency as well as on the main flux. The usual arrangement is to connect the interpole coil in series with the exciting winding, so that it carries the exciting current, which, apart from saturation, is proportional to the main flux. Exact neutralization is then only possible at one particular frequency, and the transformer voltage must be limited so that the residual voltage obtained at other frequencies does not exceed the voltage that can be dealt with by the brush contact.

So far, the reactance voltage and transformer voltage have been considered separately, but it is evident that the conditions will be modified when both are present in the same coil at the same time. For the purpose of neutralizing these voltages by means of interpole windings, however, it is quite accurate to treat them separately, as it is only the residual voltages due to incomplete neutralization which can combine together. The problem, which is more complicated because it is the same voltage, namely that at the brush contact, which has to neutralize both reactance and transformer voltage, is discussed in Section 18.

17. Damping windings

An A.C. commutator machine, particularly one of the Scherbius type, when provided with interpoles is capable of handling an output which is a reasonable fraction of that obtained with a corresponding D.C. generator. However, the majority of polyphase commutator machines are constructed so that their operation depends on the production of a rotating field as in an induction motor, and therefore interpoles cannot generally be used. The use of high resistance connectors between the armature winding and the commutator segments has been mentioned in Section 15, and considerable benefit can be obtained by this construction. The action of resistance connectors is explained more fully on p. 98. They have the disadvantage that appreciable additional losses occur, and they cannot be used if the motor is liable to be stalled, because, with the rotor stationary, one connector remains in circuit and overheats. A much greater improvement in the conditions determining commutation is obtained by the use of damping windings, so much so, that an effective damping winding can give results comparable to those obtained by using interpoles. Although a damping winding cannot compensate the fundamental reactance voltage as completely as a correctly adjusted interpole, it is effective in damping high frequency pulsations which cannot be directly neutralized by an interpole flux.

The simplest type of damping winding consists of short-circuited coils wound round each armature tooth. These coils are inductively coupled with the coils of the main winding, and absorb some of the energy of commutation. The resistance of the damping coils must be fairly high because of the losses which occur, and their effectiveness is limited by this. Better results are obtained with the so-called 'figure-of-eight' winding illustrated in Fig. 43. This damping winding also consists of loops of wire wound round each tooth, but pairs of loops round adjacent teeth are series connected in opposition to form a figure-of-eight, so that the voltages induced in the two loops by the main flux largely cancel each other. With this arrangement a much lower resistance can be used without causing excessive loss. In Fig. 43, only half the coils are shown for clearness; in practice, every pair of adjacent teeth is embraced by a coil.

The most effective types of damping winding consist of auxiliary coils wound in the armature slots so as to form an additional lap winding of the normal type, which is connected to the commutator segments in parallel with the main winding. A number of different arrangements of this type of damping winding have been used with great success. To obtain the best results a damping coil must fulfil two conditions:

(*a*) It must not be linked inductively by the leakage flux of the main coil to which it is connected.

Fig. 43. 'Figure-of-eight' damping coils

(*b*) There must be little or no circulating current due to the main flux of the machine, in the closed circuit comprising a main coil and a damping coil. Whatever the arrangement, the balance of voltages between the main and damping winding in respect of both magnitude and phase, is maintained by suitable choice of the pitch of the coils and their location round the armature.

Split coil winding for direct current machines

The forerunner of the damping windings now used in A.C. commutator machines is the split coil winding used in large D.C. machines having only one coil per slot. Each coil is split into two equal parts, one under-pitched and one over-pitched, so that each slot has four bars, two in each layer. A developed diagram of this winding is shown in Fig. 44, and for clearness one split coil is in heavy lines. When the brush ceases to short-circuit two segments, the flux in the teeth between the two half coils does not have to die away instantly because a circulating current can flow temporarily in the closed circuit formed by the two half coils.

Simplex discharge winding

In the simplex discharge winding (Fig. 45) the main coils are short-pitched, while the damping coils are over-pitched by the same amount as the main coils are under-pitched. Thus both the

Fig. 44. Split coil winding

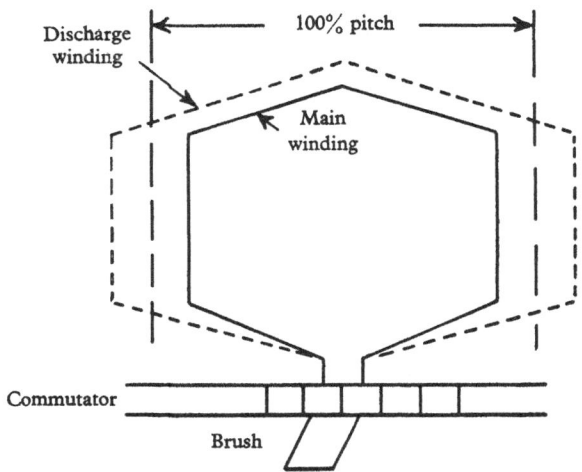

Fig. 45. Diagram of simplex discharge winding

above conditions are satisfied if the main rotating flux is sinusoidal, since the two coils lie in different slots, and the total voltage round the combined circuit is zero. The damping coil has a smaller section than the main coil and is located at the top of the slot (Fig. 46) so that its resistance is greater, and its inductance less, than those of the main coil. The complete winding thus consists of two ordinary double-layer lap windings, one above the other and connected to the same commutator lugs.

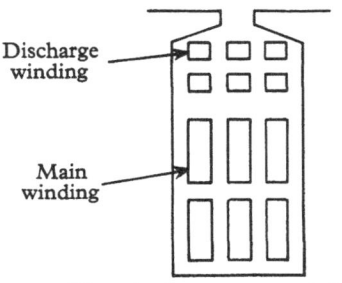

Fig. 46. Arrangement of slot with discharge winding

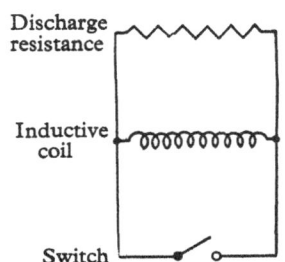

Fig. 47. Diagram of switch with discharge resistance

The action of a damping winding can best be explained by comparing it with the discharge resistance connected across a switch used to open an inductive circuit such as a generator field winding (Fig. 47). When the current in the coil is reduced from its initial value to zero by opening the switch, the flux linking the coils must also be reduced to zero if no discharge circuit is provided. This rapid change of flux induces a high voltage, which causes the switch to arc, thus tending to prolong the period during which current flows. If there is a discharge circuit, current can continue to flow in the main coil even after the switch is opened, because the induced voltage passes a current through the resistance. Hence the rate of change of flux is considerably less and the danger of sparking is reduced. Energy which would otherwise cause an arc is dissipated in the discharge resistance.

Discharge winding

Main winding

Commutator

Brush

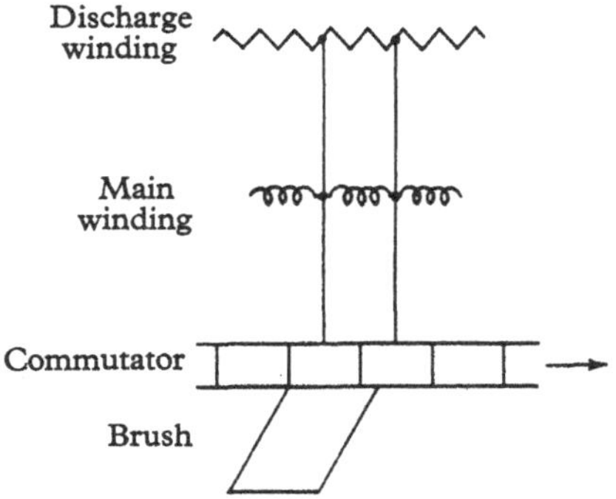

Fig. 48. Diagram of commutator with discharge winding

The same action takes place in a coil of a commutator winding as the two segments to which it is connected pass under a brush (Fig. 48). If the voltage induced in the coil, due to the change in the flux linking it, is greater than the total brush contact voltage which normally absorbs the energy of commutation, sparking will result when the circuit (which has been closed by a brush) is broken. By connecting a coil of a damping or discharge winding in parallel with the main coil, a circuit is provided in which current can continue to flow after the brush has ceased to short-circuit the segments, so that the leakage flux linking the coil need not change as rapidly as it would have to otherwise. Some of the energy of commutation is thus transferred to the discharge resistance.

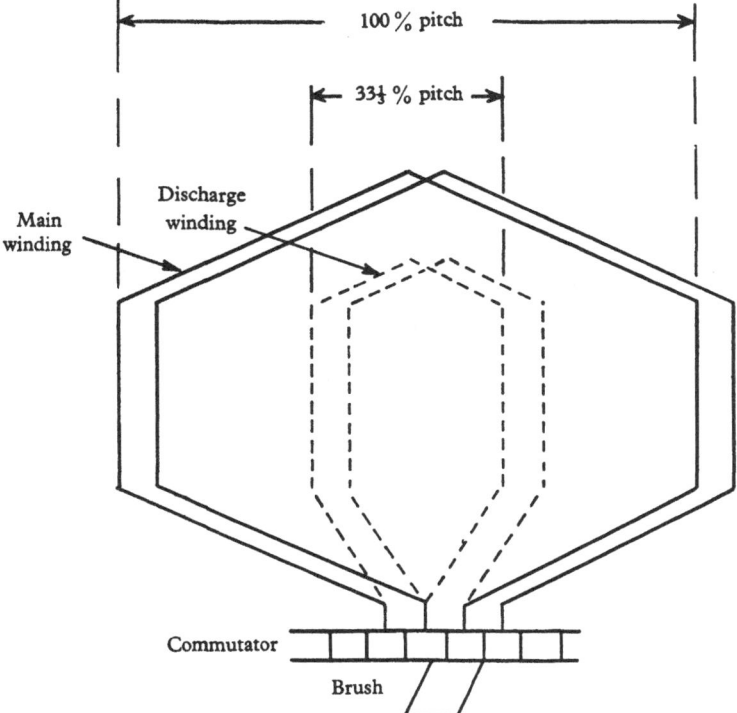

Fig. 49. Diagram of duplex discharge winding

Duplex discharge winding

The principle of the simplex discharge winding can be extended for use with the duplex type of main winding, if the pitch of the damping coil is shortened so that the voltage induced in it by the main flux is approximately half that in a main coil. Generally the pitch of the damping coil is $33\frac{1}{3}\%$ so that the pitch factor ($=\sin 30°$) is 0·5, as in Fig. 49. The voltage in two damping turns is then balanced by that in one main coil. In addition to its action as a discharge circuit, the damping coil also performs the important function of holding the potential of each segment intermediate between the potential of the two adjacent segments with which there is no connexion through the main duplex winding. Thus the back-to-front connectors described on p. 35 can be omitted.

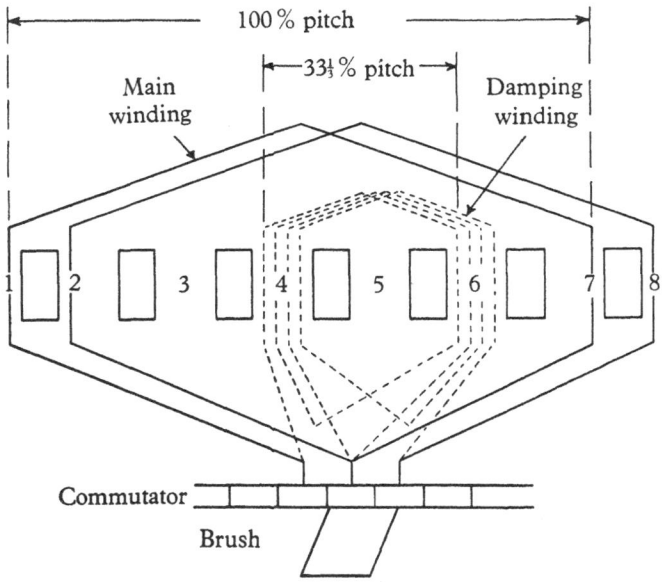

Fig. 50. Diagram of transformer type damping winding

There are two main variations of the duplex discharge winding.

(1) *Full-pitch main winding*. The damping winding used with a full-pitch main winding is a straightforward lap winding with $33\frac{1}{3}\%$ pitch. In the main duplex winding, one coil out of each group of coils associated with a slot must be over-pitched by one slot as in Fig. 25; the remainder have full pitch. It can be shown that complete equalization of voltages is obtained with this arrangement.

(2) *Short-pitch main winding*. Many alternative arrangements are possible when the main winding is short pitched. Sometimes exact equalization of voltage can be obtained, while in others some out-of-balance, resulting in circulating currents and corresponding losses, has to be tolerated. Generally the auxiliary winding has $33\frac{1}{3}\%$ pitch, but alternate coils must be displaced relative to the remainder so that the voltage vectors add together in a similar way to those of the two sides of a short-pitched coil of the main winding.

Transformer type damping winding

The principle involved in the transformer type damping winding (Fig. 50) is to link each main coil with other adjacent main coils by transformer action. It is used with a simplex main winding, and each damping coil connected between adjacent segments consists of two turns having $33\frac{1}{3}\%$ pitch. The damping winding is located at the bottom of the slots in a separate compartment which is separated from the upper part by steel shims as shown in Fig. 51. Each pair of double damping coils constitutes a transformer which couples a main coil to an adjacent main coil and consequently reduces its effective inductance during commutation.

Each damping coil has its axis

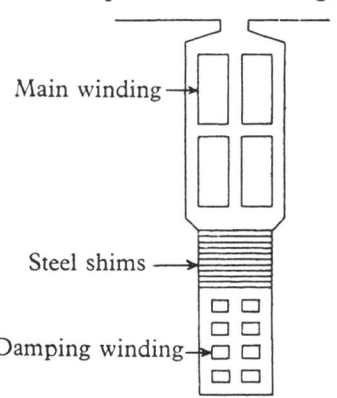

Fig. 51. Arrangement of slot with transformer type damper winding

displaced relative to the axis of the main coil to which it is connected, and hence it is not possible to equalize the voltages exactly with the transformer type of damping winding. Therefore, the resistance of the damping coil must be made high enough to ensure that the circulating currents are not excessive.

To cater for large machines with high values of flux, the winding may be modified by introducing intermediate segments between the segments connected to the main winding, and connecting these segments to intermediate points in the damping winding. The results obtained in this way are similar to those obtained with the duplex windings already mentioned.

More detailed information on these damping windings can be obtained from the technical articles listed in the bibliography at the end of the book.

18. The action of the brush in producing good commutation

Brush contact voltage

It has been shown that the difficulties associated with the collection of alternating current from a commutator can be expressed in terms of two voltages, the reactance voltage and the transformer voltage, set up in the coils undergoing commutation. When the machine is not provided with interpoles or damping windings the whole of the reactance and transformer voltages have to be absorbed in the coil or at the brush contact, but, if interpoles are provided, only smaller residual voltages have to be absorbed. The circuit in which the voltages act consists of one or more coils with a segment at each end, two brush contacts, and the brush itself. The voltages available for neutralizing the commutation voltages are the resistance drops in the coils and connecting strips, and the voltage drop at the brush contact. The resistance of the brush itself can be neglected.

The brush contact voltage, which is of fundamental importance in determining commutation, depends on a number of factors, such as the quality of the brush material, the pressure between brush and commutator and the current density at the contact surface. For steady conditions in a given machine, a curve may be drawn showing the relation between the voltage drop and the current density. The brush contact is equivalent to a resistance whose value varies with the current. Fig. 52 shows a typical brush drop curve for the electrographitic or bakelite bonded types of carbon brush commonly used on A.C. commutator machines. The mean current density due to the main current of the machine is generally not greater than 60 amp./sq. in. of brush contact surface. This current density corresponds to a voltage drop of 1·75 V. between brush and commutator and the energy loss at the brush contact due to the main current is therefore 105 W./sq. in. of brush surface with this value of current density.

Distribution of voltage drop across brush

The curve shown in Fig. 52 is determined by measuring the voltage between a slip-ring and a brush carrying direct current, under conditions where the current is uniformly distributed over the contact area. But when the brush is on a commutator and circulating currents flow in the coils short-circuited by the brushes, the distribution of current over the area of the brush is not necessarily uniform at any instant. At any point, however, the same

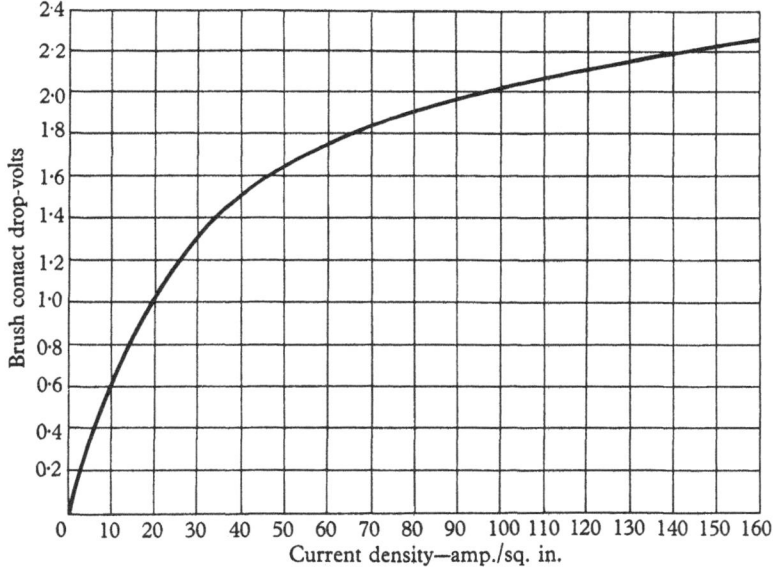

Fig. 52. Typical brush contact drop curve

relation holds between voltage drop and current density and a curve may be plotted showing the variation of current density or of voltage drop across the thickness of the brush. Under ideal conditions the current density and voltage will be the same at all points on the same line parallel to the axis of the machine.

Fig. 53 is a curve showing the relation between the current density at any point and the distance of that point from the edge of the brush, in the simple case where the current density varies linearly with the distance between the edges of the brush. A curve similar to this is obtained in a D.C. machine without interpoles.

The *leading edge* of the brush is that which meets the oncoming segments first, while the *trailing edge* is that which breaks contact with the segment when commutation is complete. It can be seen that the current distribution shown in Fig. 53 is equivalent to a uniform current density represented by the line *AB*, with a

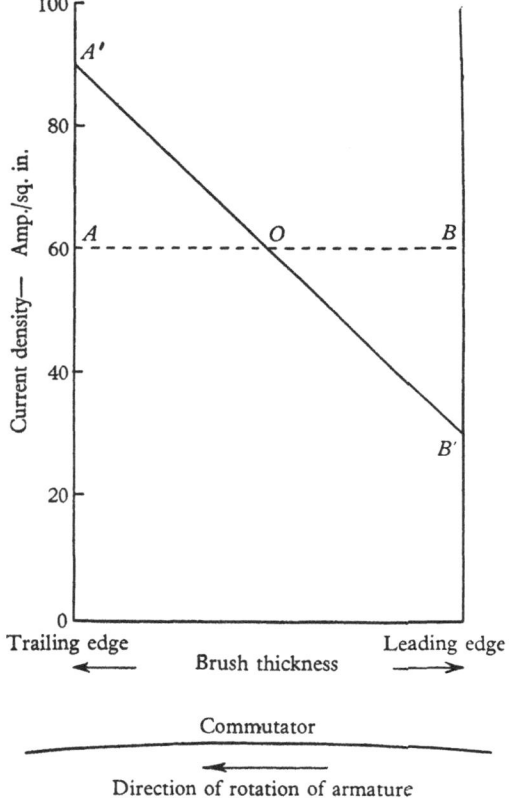

Fig. 53. Distribution of current density at brush contact

circulating current superimposed represented by the positive area *AOA'* and the negative area *BOB'*. The value of voltage drop corresponding to the current density at each point may be found from Fig. 52 and a curve of voltage drop plotted as in Fig. 54.

In any machine the voltage and current distribution over the thickness of the brush at any instant may be indicated by two curves similar to Figs. 53 and 54. High frequency pulsations of current density at any point are accompanied by a corresponding ripple on the voltage wave. In an A.C. machine, if the current density at any point varies sinusoidally, the voltage drop at that point varies periodically in a manner determined by the curve connecting voltage and current density. An equivalent sine curve can be found for this voltage.

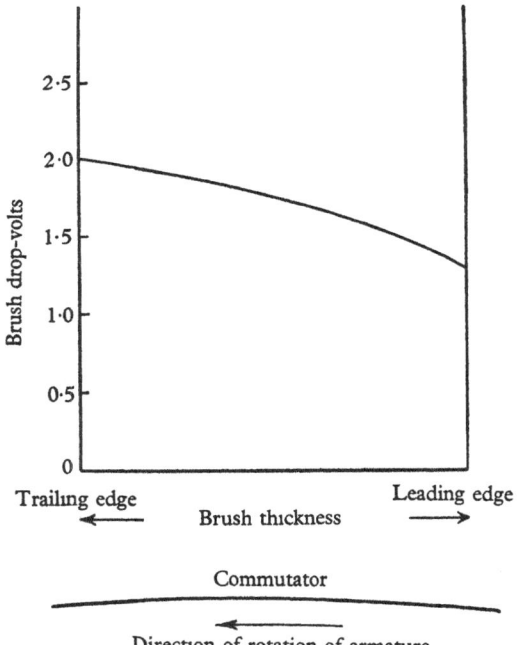

Fig. 54. Distribution of brush volts at brush contact

Measurement of brush volts

An approximate measurement of the *brush-volt curve* on an actual machine may be made by using brush pencils which are held fixed against the rotating commutator on the same axial line as the points on the brushes at which the voltage drop is to be determined. The measured voltage between the brush pencil in any position and the brushes gives one point on the brush-volt curve. This test is a common one on D.C. machines, and is also often carried out on A.C. machines of the Scherbius type using direct current, so as to test the current commutation under conditions when the transformer voltage does not exist. Usually three readings are taken—at the two edges and at the middle of the brush. The difference between the voltage drops at the opposite edges of the brush gives an indication of the amount of the reactance voltage which is being absorbed at the brush contact. The method does not record the high frequency pulsations and the readings obtained are mean values. The curve of mean current density may be derived from the curve in Fig. 52.

Readings of brush volts also may be obtained in the same way with an A.C. voltmeter when the brush is carrying alternating current, but the results are not so easy to interpret because the phase of the voltage will not be the same at different points. It is more satisfactory to use two brush pencils bearing on the commutator at points in line with the opposite edges of the brush, and to measure the voltage between them. This voltage gives an indication of that part of the total commutation voltage—reactance and transformer voltages combined—which is absorbed at the brush contact. The oscillogram in Fig. 55 was taken in this way and shows the high frequency ripple superimposed on a low frequency alternating voltage wave. A voltmeter connected between the two pencils indicates the virtual value, which is the sum of the residual reactance and transformer voltages.

Reactance voltage without transformer voltage

The resulting commutation in a machine is due to the combined and simultaneous action of both reactance and transformer voltages, but, in order to understand what happens during commutation in any coil, it is clearest to consider first the cases where one voltage is present without the other. The case where reactance voltage exists without transformer voltage will first be examined for a machine which has no interpole windings, and in which the brush is only thick enough to short-circuit two segments at a time. This arrangement is illustrated in Fig. 56 for three successive instants, during which one segment is moving past the brush. Current is flowing into the brush and the commutator is moving towards the left. During the very short period of commutation the current in the brush, the frequency of which is not generally more than 50 cyc., can be assumed to remain constant,

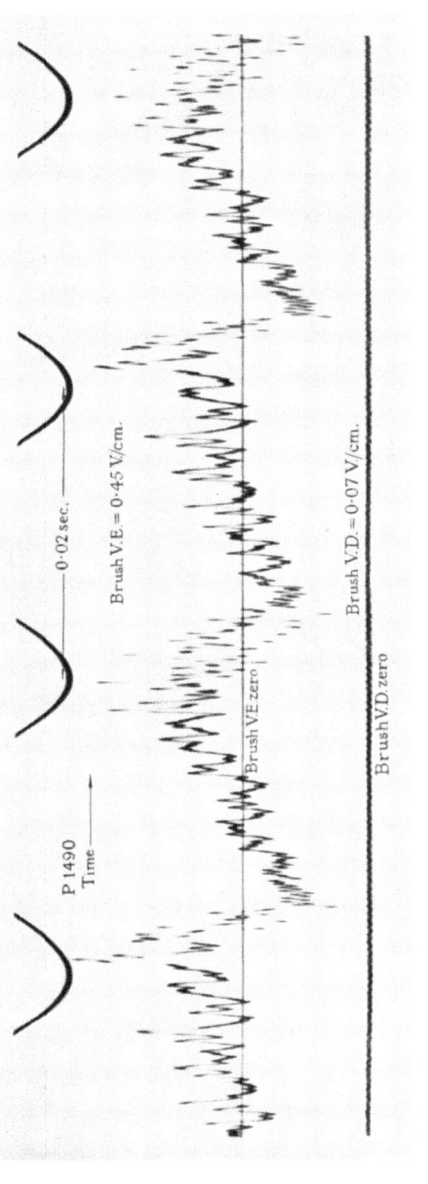

P 1490
Time

0·02 sec.

Brush V.E = 0·45 V/cm.

Brush V.E zero

Brush V.D. = 0·07 V/cm.

Brush V.D zero

Fig. 55

to face p. 90

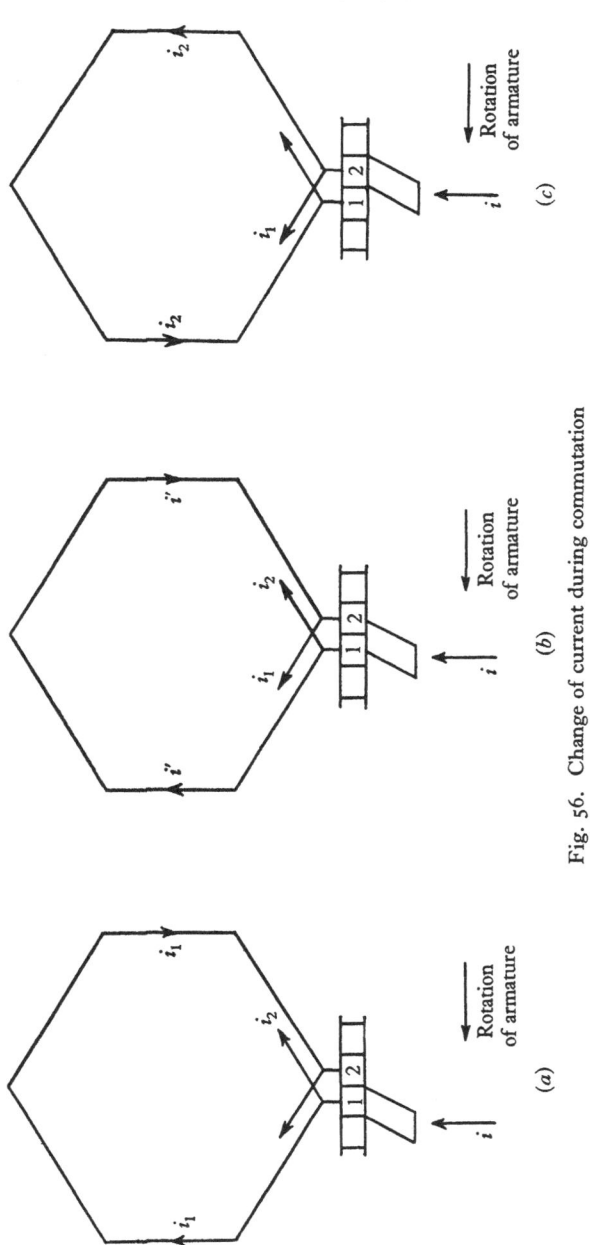

Fig. 56. Change of current during commutation

so that the argument applies equally for alternating or direct current, if we assume that no transformer voltage exists.

The current i flowing into a brush is divided into currents i_1 and i_2 flowing in opposite directions through the winding. When the brush is in the position shown in Fig. 56(a), the current i_1 flows round the coil shown in a clockwise direction from segment 1 to segment 2. A short time later when the brush is in the position shown in Fig. 56(c), the current in the same coil is i_2 and flows in a counter-clockwise direction. The total change in the current in the coil is $i_1 + i_2$, and is equal to the current flowing into the brush, as stated on p. 68. In each of these positions the current is uniformly distributed over the contact area between the brush and the segment on which it is resting. In any intermediate position such as that shown in Fig. 56(b), when the brush short-circuits the two segments 1 and 2, the current in the coil is changing and has a value i'—taken in a clockwise direction— intermediate between the initial positive value and the final negative value. If, as is assumed, no other voltage is introduced into the coil by transformer action or by an interpole flux, the rate of change of current in the coil is determined solely by the resistance drop in the coil and connecting strip, the inductance of the coil, and the algebraic sum of the contact drops between the two segments and the two parts of the brush.

As soon as the brush makes contact with segment 2 current flows between them, thus changing the current in the coil. If the change in current took place at a uniform rate as the segments moved under the brush, the current density would remain uniform over the whole contact area and there would be no difference in voltage between the two segments. The change in current can, however, only take place if a voltage is present in the circuit opposing the inductive voltage due to the change in current, that is, the reactance voltage. The change in current therefore lags so that the current density between segment 1 and the brush is greater than that between segment 2 and the brush and the difference in brush-contact voltage is used to neutralize the reactance voltage. This process continues until all the current is flowing through segment 2 and none through segment 1.

The change of current is accelerated at the end of the period of commutation because the area of contact between the brush and segment 1 is decreasing rapidly with a consequent increase in current density and voltage drop. Even at very high densities, however, the available brush drop is limited and, unless the reactance voltage is kept at a low value, it is not possible to complete the change of current before the circuit is broken, and sparking results.

The simple case described has been chosen for the sake of clearness. In practice the brush is usually thick enough to short-circuit more than one coil, that is, the thickness of the brush is greater than the combined thickness of one segment and two pieces of insulation. The advantage of this is that a coil undergoing commutation is linked inductively with another coil which is short-circuited by the brush before the first coil is open-circuited. It is evident that the equivalent self-inductance of a coil is reduced if it is closely linked with another coil which is short-circuited on itself, because a change of current in the first coil calls forth a current in the short-circuited coil which tends to neutralize the flux produced by the first coil. In other words, the energy in the field surrounding a coil undergoing commutation, instead of being dissipated in sparking when the circuit is broken, is transferred to the next coil. The short-circuited coil acts temporarily as a simple damping winding.

In order to obtain the full advantage of using a thicker brush, the number of segments per slot should be two or more and the pitch of coils in different parts of the slot should be different. Fig. 57 shows three coils of a winding of this type, which is called a *stepped winding*. The three coils A, B and C each have one side together in the slot on the left-hand side of the diagram, but the right-hand side of coil A is in one slot while those of coils B and C are in another. When the segments are approaching the position shown relative to the brush, that is, the point at which coil A becomes open-circuited, the coil B, which is inductively coupled in the left-hand slot with the coil A, is short-circuited by the brush. In the same way every coil is inductively coupled at least on one side with the coil following. This is not the case with the winding shown in Fig. 40 in which all the coils have the same pitch. The effect of this mutual inductance between one coil and the next is to improve the conditions during the final part of the period of commutation of a coil and to prevent sparking. As stated above, the energy stored in the field surrounding the slot, which would otherwise be dissipated in the form of a spark at the trailing tip of the brush, is now passed on to the next coil.

Although the commutation can be considerably improved by the use of a stepped winding, it is evident that the reversal of current must be almost completed before the final part of the period of commutation. That is to say, the reactance voltage calculated on the assumption that the change in current is uniform, must not be much greater than the difference in brush-contact drop which causes the current to change. When the brush

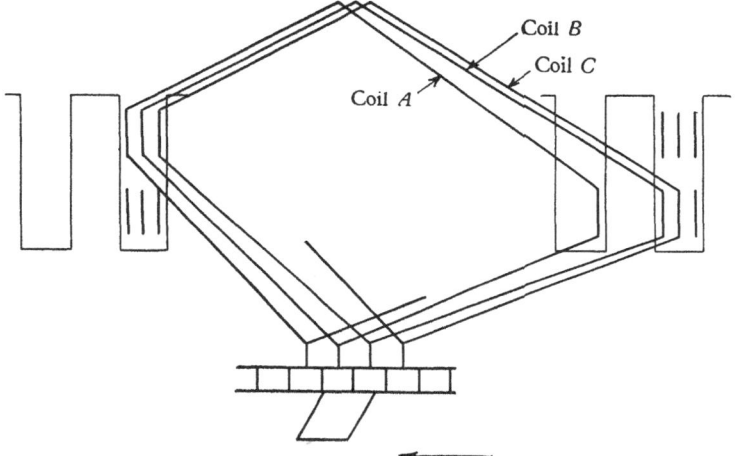

Fig. 57. Stepped winding

short-circuits more than one coil at a time the total reactance voltage across the thickness of the brush is greater than the reactance voltage per coil E_c calculated by equation (32) in proportion to the number of segments covered by a brush. In A.C. commutator machines of the types which are not provided with interpoles, the brush thickness is usually less than twice, but more than once, the pitch of a segment. With this arrangement two coils are short-circuited during part of the period of commutation, and only one coil during the other part, and the mean reactance voltage across the brush is more than E_c but less than $2E_c$. In the general case the mean reactance voltage across the brush face is taken as $b/c \times E_c$, where b is the brush thickness and c the pitch of the segments. It is found in practice that, in machines without commutating poles or damping windings, the total reactance voltage calculated in this way must not exceed about 0·7 to 0·8 V. on the commutator between the leading and trailing tips of a brush. It can be seen from an inspection of Fig. 52, that a difference of 0·8 V. in contact drop between different parts of the brush corresponds to a considerable variation of current density. When a damping winding is used, a considerably greater value of reactance voltage than this is permissible.

Transformer voltage without reactance voltage

The transformer voltage induced in the coil short-circuited by a brush depends on external factors such as flux and frequency, and not on the brush current. It thus differs fundamentally in its origin from the reactance voltage which arises from the change of current in the coil. In dealing with the transformer voltage, the problem is simply that of absorbing what is equivalent to an externally applied voltage in a circuit in which the brush contacts constitute a resistance. Consider first the condition where there is no main current, so that the reactance voltage is zero.

Assuming, as before, that the brush is thicker than one segment but not so thick as two segments, two alternative positions are taken up in turn. In the position shown in Fig. 58(a), only one coil is short-circuited by the brush. Assuming the main current flowing through the brush to be zero, a circulating current will flow across the two sections of the brush contact in opposite directions between brush and commutator, but in the same direction round the short-circuit as indicated by the arrows. If the current attained a steady value the transformer voltage in the coil would be balanced against the sum of the two brush-contact drops and the ohmic resistance of two connecting strips and one coil. In the second position shown in Fig. 58(b), if the same assumptions are made the transformer voltage in two coils is balanced against two brush-contact drops plus the ohmic resistance of two connecting strips and two coils, in the circuit indicated by the arrows.

If the coils of the winding are connected to the commutator segments without any intermediate high resistance connectors, the ohmic resistance drop in the circuit is usually small compared with the voltage drop at the brush contact. The permissible transformer voltage is therefore determined, in accordance with the brush-voltage curve, Fig. 52, by the permissible current density at the part of the brush contact included in the short-circuit. The final determining factor is the loss at the brush contact—known as the *parasitic loss*—which affects the temperature of the commutator and brushes, as well as the efficiency of the machine. If the parasitic loss is excessive the heat at the brush contact disintegrates the surface and rapid brush wear results.

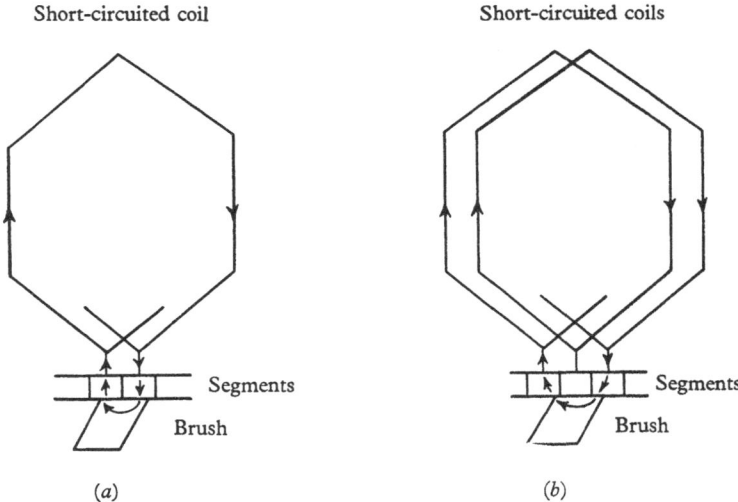

Fig. 58. Circulating currents in short-circuited coils

With the assumptions made above that the main brush current is zero and that the circulating current is determined only by the transformer voltage and the resistance of the circuit including the brush contact, the current density at the brush surface and the parasitic loss can be calculated for any position of the brush relative to the segments. The simplest cases are those in which the brush is placed symmetrically relative to the segments as shown in Fig. 58(*a*) and (*b*). In Fig. 58(*a*), two brush contacts are included in the circuit each having a thickness of $b/2$, if the thickness of the insulation is neglected. The total voltage drop round the circuit can be calculated for any current, by adding the resistance drop to twice the voltage given by the brush-drop curve (Fig. 52) corresponding to the current density, which is assumed uniform over each part of the brush. The current which flows when a voltage E_t is applied can therefore easily be deduced by plotting a curve or by trial and error, and this current multiplied by E_t is the loss in the circuit. A similar calculation can also be made for the position shown in Fig. 58(b), where the voltage is $2E_t$ and the thickness of each part of the brush contact in the circuit is $(b-c)/2$, neglecting the thickness of the insulation.

As an approximation, it may be assumed that the value P_1 of the power loss for the position in Fig. 58(a) is the mean loss during that fraction of the total time taken to pass a segment, during which one coil only is short-circuited—namely, $(2c-b)/c$ of this period—and that P_2, the loss calculated for the position in Fig. 58(b), is the mean loss during the remainder of the period when two coils are short-circuited—namely, $(b-c)/c$ of the period. Hence the mean power loss over the whole period is

$$\frac{2c-b}{c}P_1 + \frac{b-c}{c}P_2.$$

The period under consideration is very short compared with the A.C. cycle, so that the value of parasitic loss given by this expression corresponds to an instantaneous value of the transformer voltage. If the voltage is assumed to vary sinusoidally with time a curve of parasitic loss against time could be plotted. Owing to the shape of the brush-voltage curve the parasitic loss increases rapidly with the voltage, and the mean parasitic loss over a cycle is therefore much greater than the value corresponding to the virtual value of voltage. On the other hand, the parasitic loss is actually much less than would be calculated by the method indicated, because this assumes steady conditions, and neglects the inductance of the coil. The current in the short-circuited coil does not, in fact, ever attain the steady value. The actual value of the parasitic loss in a machine is very difficult to predetermine and can, in practice, vary considerably between machines of similar design, owing to small variations of brush bedding, or differences in the mechanical running.

The parasitic loss may be measured on test by running the machine light, that is, with no main current in the brushes, and measuring the input, first with brushes raised, and then with brushes down. The difference between the two input figures includes the brush-friction loss, which has to be determined on a separate test, and deducted. The effect of the main brush current, which flows when the machine is loaded, is to modify the current distribution under the brush by adding at every point a current density which is uniform over the brush. It is usually assumed when calculating efficiencies that the total loss at the brush contact is equal to the sum of the parasitic loss measured when the main current is zero, and the direct contact loss calculated from the brush-voltage curve on the assumption that the parasitic loss is zero. This method overestimates the loss on load because the parasitic currents do not have a separate existence from the main currents. This would not affect the accuracy of the results if the brush-contact resistance were constant. But because the contact resistance decreases as the current density increases, the loss occurring in practice is less than the simple sum of the two sets of losses calculated separately.

A common value for the transformer voltage per segment for the condition illustrated in Fig. 58 is 2·0 V. The peak value of the A.C. wave is 2·83 V., and on the brush-drop curve this voltage corresponds to a current density of several hundred amp./sq. in. Very high current densities of this order may actually occur in the position in Fig. 58(b) when the voltage in two coils is absorbed by two brush contacts, but if the area of the part of the brush carrying this current density is only a fraction of the whole, and if also the time during which this current flows is only a fraction of the total period during which the segment passes under the brush, then the mean parasitic loss which results may still be small. This shows the importance of using brushes which are as thin as possible provided they are thick enough to touch three segments for a short period. No advantage is gained by making the brush thinner than a segment, as the theoretical reduction of parasitic loss is offset by the fact that the energy in the field surrounding the short-circuited conductors can no longer be transferred to the next coil.

In order to obtain larger outputs, A.C. commutator machines are frequently provided with high resistance connectors between the armature winding and the commutator. These resistance connectors are usually made of German silver wire and may have as much as twenty times the resistance of a coil. Any particular connector is only included in the main circuit intermittently, but two connectors are always included in the commutation circuit in which the transformer voltage is absorbed. By this means transformer voltages of 2·6 V. per segment or more can be successfully dealt with, as previously mentioned.

Although the inductance of the coil is useful in keeping down the mean circulating current due to the transformer voltage, it also has the effect of causing sparking at the trailing edge of the brush when the circuit carrying the circulating current is opened. This provides an additional reason for using the 'stepped' type of armature winding. Unless the parasitic loss is excessive, however, it is found that this sparking does not cause destructive wear and should not cause the concern that similar sparking in a D.C. machine would give.

Comparisons between action of reactance and transformer voltages

It should be clear from the foregoing why the permissible value of reactance voltage is much less than the permissible value of transformer voltage. The reason is that the variation of current density over the thickness of the brush is limited to a much smaller amount when the limit is imposed by the rate of change of the current in the coil than when it is imposed only by the loss occasioned by circulating currents. For the same reason, resistance connectors are of much greater assistance in dealing with voltage commutation than with current commutation, because the circulating current, and hence the ohmic drop in the connectors, is much greater.

The action of interpoles in assisting commutation is also different in the two cases. It is evident that the voltage induced in the short-circuited coil by rotation in the interpole flux, is of the same character as the transformer voltage, that is to say, it is introduced into the circuit by external causes. When an interpole winding is used to neutralize the transformer voltage, the total induced voltage is reduced to a value equal to the vector difference of the transformer voltage and the voltage due to the interpole flux. This residual voltage then sets up circulating currents in the way described and is absorbed by the voltages at the brush contact, and in the coil and connectors.

The action of the interpole in assisting current commutation is different. It is the same as that which takes place in D.C. machines

provided with interpoles. If the interpole flux has the correct
value, the voltage induced by it in the short-circuited coil enables
the current in the coil to change at a uniform rate, so that the
current density remains uniform over the brush face at the
different stages of commutation. If the interpole ampere-turns are
not sufficient, the current density varies so as to produce a voltage
difference across the brush, but the value of this voltage difference
will be less than it would be if no interpole were provided. If, on
the other hand, the machine is over-compensated on the interpole, there is an excess of induced voltage over the value required
to cause the change of current, and this residual voltage sets up
circulating currents in the commutation circuit in the same way
as a transformer voltage induced by the alternation of the main
flux. The comparison between over-compensation and under-
compensation on the interpole is therefore similar to the comparison between the conditions caused by transformer voltage
and by reactance voltage. This is one reason why it is better to
keep the interpole ampere-turns on the high side, that is, to over-
compensate rather than under-compensate.

Combined action of reactance and transformer voltages

It has already been explained that the problem of commutation
in an A.C. commutator machine, which has been discussed only
for the cases in which the reactance or transformer voltages exist
independently, is complicated by the fact that the two voltages
generally exist together. A very rough idea of the combined effect
can be obtained by adding together the two voltages vectorially,
but this is quite inaccurate as a quantitative guide, because of the
different character of the two voltages. The best analysis is obtained
by resolving the transformer voltage into two components, one in
quadrature with the current and one in phase. The quadrature
component is zero when the reactance voltage is maximum, and
vice versa, and these two voltages act more or less independently
of each other. The in-phase component of transformer voltage, if
it opposes the reactance voltage, is equivalent to the voltage induced due to an interpole used for assisting current commutation,
but if it is in the same direction as the reactance voltage, it has the
effect of making the commutation considerably worse than it
would be if either of the two voltages were present by itself.

Consider as an example a machine with a uniform air-gap having an armature winding similar to that illustrated in Fig. 39. The reactance voltage is always in phase with the current in the brush. The transformer voltage on the other hand is related in phase to the main flux and hence to the induced voltage. Neutralization of the reactance voltage could be carried out by providing a compensating winding with more ampere-turns than the armature, thus producing a flux due to the main current opposite in space phase to the armature M.M.F. If, therefore, the main flux, set up in any way, has this same direction, the transformer voltage induced by it in the short-circuited coil will oppose the reactance voltage. Hence, from the point of view of commutation, it is preferable to make the component of the main rotating flux which is in phase with the armature M.M.F., opposite in direction to that M.M.F.

It was shown on p. 47, that a flux which is opposite in direction to the armature M.M.F. induces a voltage in the armature winding which is in quadrature with the current, and which lags behind the current if the flux rotates in the opposite direction to the armature. Hence if the main current and voltage are in phase, the transformer voltage in a short-circuited coil is in quadrature with the reactance voltage and also in quadrature with the main voltage of the corresponding phase. Thus, if the voltage induced in the commutator winding is in phase with the current, as for example in a machine used for regulating the speed of an induction motor without affecting the power factor, the transformer and reactance voltages are at right angles to each other and will act more or less independently of each other. Under this condition, if both voltages are limited to values which give satisfactory commutation when acting separately, the commutation will be only slightly less satisfactory when they are present together.

Usually, however, the current is made to lead the voltage in order to compensate for the inductive effect of other windings or apparatus in the circuit, as in a machine used for producing power-factor correction as well as speed regulation in an induction motor. In a phase advancer, for example, the voltage is approximately at right angles to the current. As shown above, the component of voltage at right angles to the current lags behind the current so as to produce power-factor correction and at the same time the associated component of transformer voltage in the short-circuited coils opposes the reactance voltage, if the rotating flux is opposite in direction to the armature M.M.F., and rotates in the opposite direction to the armature. There is often a very noticeable difference in the commutation of an A.C. commutator machine, if the sequence of the phases is reversed without changing the relation between voltage and current. The difference between the commutation at the two sets of brushes in machines such as the Schrage motor, where the same current passes into one brush and out at another is attributable in part to the fact that the direction of the transformer voltage relative to the reactance voltage is opposite in the two cases.

In the salient pole machine, similar considerations apply, but the relative direction of transformer and reactance voltages is different sometimes from that in a machine with a purely rotating flux. It is still true that the transformer voltage is in quadrature with the induced voltage in the corresponding phase, but the different distribution of the main flux introduces other factors with regard to the actual direction of the transformer voltage.

The explanations given in this section show how extremely important the brush is in obtaining good commutation. Considerable advance has been made since the early days of A.C. commutator machines in improving the characteristic of the brush, that is, in raising the voltage drop at high current densities without increasing it seriously at low densities. Improvements in this respect are still being made, as well as in the mechanical properties of the brush. The ideal brush would be one whose characteristic was approximately a straight line through the origin, since the direct contact loss would then vary on load and on overload in the same way as the resistance loss in a winding, but high values of transformer voltage could be dealt with without causing excessive parasitic loss.

19. Comparisons with direct current machines

The object of the discussion on commutation in this chapter is to show the ways in which A.C. commutator machines differ in this respect from the better known D.C. machines. There are many points of importance which apply equally to both types of machine, but which need not be considered in detail here. In particular, commutation depends greatly on the design of the brush-holder, the smooth running of the commutator, and the uniform quality of the brush carbon. Accurate spacing of both segments and brushes round the periphery of the commutator is of even greater importance in A.C. than in D.C. machines.

It has been shown that commutation is more difficult in A.C. than in D.C. machines for two reasons. In the first place commutating poles cannot be as easily provided, and when used cannot be as easily adjusted. Secondly, the transformer voltage of commutation, which does not exist in a D.C. machine, has to be absorbed in the coil short-circuited by the brushes. There are, however, other factors which act in favour of the A.C. machine.

In A.C. machines where the speed of an induction motor is regulated by injecting the voltage from a commutator winding into the secondary circuit, the power carried by the commutator is only the slip power corresponding to the percentage regulation from synchronous speed, and is less than the output of the motor. The commutator may for this reason be no larger than that of a D.C. motor delivering the same power as the induction motor in spite of the difficulties just mentioned.

Another advantage of A.C. commutator machines over D.C. machines is that the same amount of sparking at the brushes is less destructive if the current alternates. It is usual to specify that a D.C. machine must operate at full load without visible sparking, and even then it is important to provide a considerable wearing depth on the commutator bars to ensure a reasonable length of life. In an A.C. commutator machine, on the contrary, there may often be a certain amount of visible sparking without harm to either commutator or brushes.

The reason for this is to be found in the nature of the brush-contact voltage. The voltage drop is set up by a kind of electrolytic action, and particles of matter are transferred from the anode to the cathode. It is an experimental fact in a D.C. machine, that copper is carried from the commutator to the brush if the current flows from the commutator to the brush, and that carbon is transferred from brush to commutator if the current flows in the opposite direction. When the current alternates, this action cannot take place. Experience has shown that A.C. machines will often operate for long periods without requiring attention to the brushes or commutator, even though there is a degree of visible sparking which could not be tolerated in a D.C. machine.

Nevertheless, it is important to ensure that the quality of commutation is within the limits found in practice to be necessary to obtain satisfactory service. Much experience is needed to determine these limits and to know the details in the arrangement of the windings and magnetic circuit which give the best results. In particular, high frequency pulsations must be reduced as far as possible, and all the measures used to reduce harmonic fluxes in induction motors, such as short-pitching of the windings and skewing of the core, can be used with good effect both in the commutator machines themselves and in induction motors associated with them. The use of damping windings, especially in Schrage motors and shunt motors, has also been an important factor in extending the limits up to which an A.C. commutator machine can be used.

Chapter 5

SOME SPECIAL PROBLEMS

20. Generator operation of induction motors

It was explained on p. 11, that an ordinary induction motor, connected to a supply system, and driven by mechanical means above synchronous speed, generates power and feeds it to the supply system. An induction motor with a short-circuited rotor can only generate when it is connected to an external system which supplies the magnetizing current. If, however, a suitable voltage is injected into the secondary circuit from a commutator machine, the speed and wattless current, when the induction machine is operating as a generator, can be modified in the same way as when it operates as a motor. An induction motor with an injected voltage can be made to operate as a generator at any speed, and can, moreover, supply wattless power to, instead of taking it from, an external system.

The vector diagrams which express the relation between the currents and voltages in the machine windings still apply with suitable changes of sign. The change from motoring to generating operation is perfectly continuous and both the current locus diagram and the torque-speed characteristic can be extended beyond the axis into the generating region, just as in the case of the motor with short-circuited rotor whose characteristics are shown in Figs. 10 and 11. Any motor which drops in speed between no-load and full-load can be made to generate by raising the speed above the no-load value.

Independent generator operation of induction motors with commutator machines.

Unlike the ordinary induction motor, an induction motor with a commutator machine in the secondary circuit can, under certain circumstances, generate a voltage at a definite frequency without being tied to any supply system. That this is so can be seen from the following simple consideration. If an induction motor, supplied at constant voltage and frequency, is mechanically driven and regulated by means of a commutator machine in such a way that the current taken from the supply is zero, then the supply can be removed without making any difference. In general, as will be shown later, the motor continues in a stable condition and generates the same voltage and frequency, the losses being supplied by the driving machine. If the same conditions are fulfilled when the induction motor is not mechanically driven, the motor falls in speed slowly because of friction losses, but the same generator action takes place at first while the speed is maintained. If the primary current, before disconnecting the supply, is not zero, there may be another value of primary voltage and frequency which, with a different degree of saturation in the induction machine, would give zero current. If so, the machine will generate, when the supply is removed, that voltage and frequency which would have caused zero current to flow.

Separate excitation

The injected voltage may be obtained from a source which is independent of the primary voltage of the motor. The low frequency exciting current passed into the secondary is then analogous to the direct current in the field of a synchronous motor. The generated voltage and frequency are fixed by the magnitude and frequency of the injected voltage and the speed of rotation of the machine. The separate excitation can be obtained from a synchronous generator for example, or from a commutator frequency changer driven and supplied independently of the induction motor. Separate excitation is, however, a condition which does not often occur in practice.

Self-excitation

More often the slip-frequency voltage applied to the secondary circuit itself depends on the generated primary voltage, and a condition of self-excitation exists. The excitation of the induction motor and the voltage in its primary are then interdependent in the same way as in a shunt-excited D.C. generator. A stable voltage is generated of such a value that the secondary exciting voltage dependent on it is in magnitude, phase and frequency, exactly that required to maintain the voltage in the primary. The terms primary and secondary refer to the circuits which would be primary and secondary during normal induction motor operation.

The simplest example of this type of self-excitation is that which occurs in an induction motor whose secondary is connected to a commutator frequency changer, as illustrated in Fig. 35. The same action can take place in a Schrage motor (Fig. 36) or in an induction motor with a Scherbius machine (Fig. 38) whose main excitation is derived from a frequency changer. Conditions leading to self-excitation can also exist in the polyphase shunt motor (Fig. 33).

Conditions for self-excitation at no-load

When the induction generator is not supplying any external load the primary current is only that required to supply the excitation to the secondary circuit, as for example through the frequency changer, and can be neglected for present purposes. The flux required to induce the primary voltage is set up by a magnetizing current flowing in the secondary circuit. At the same time, a voltage is induced in the secondary, which is approximately in quadrature with the secondary current, and whose magnitude depends on the magnitude of the flux and on the relative speed between the flux and the rotor of the induction motor. As in any induction motor, the injected voltage is the vector difference of the secondary induced voltage and the impedance drop. This fact determines the voltage and frequency at which stable self-excitation takes place.

The vector diagram of secondary voltages for an induction motor with a frequency changer when operating at no-load, is

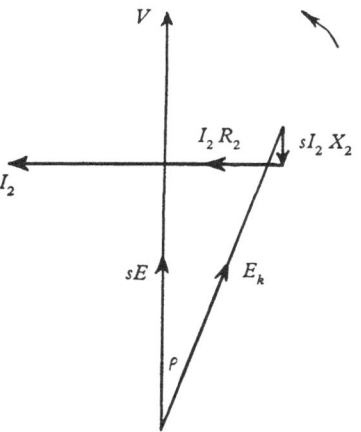

Fig. 59. Secondary voltages of self-exciting induction motor with frequency changer, on no-load

shown in Fig. 59. The secondary voltages are referred to the primary and can be represented by vectors at primary frequency as explained on p. 6. sE is the secondary induced voltage at slip s, and E_k is the injected voltage, which is proportional to, and at a constant angle to, the primary voltage. If the primary impedance drop is neglected V is in phase with E. The angle ρ between the injected voltage and the induced voltage is therefore determined by the arrangement of the combination, and is constant, as explained on p. 142. R_2 and sX_2 are the total resistance and reactance of the secondary circuit. The following two equations, in which the quantities are the scalar values, then hold:

$$E_k \sin \rho = I_2 R_2, \tag{34}$$

$$E_k \cos \rho = s(E + I_2 X_2). \tag{35}$$

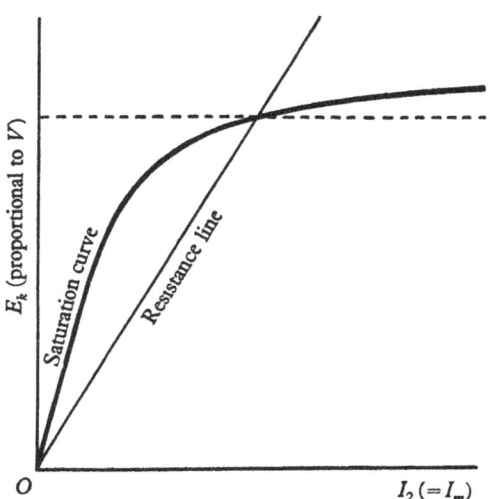

Fig. 60. Self-excitation voltage of induction motor
with frequency changer

A further relation between E_k and I_2 is given by the saturation curve of the induction motor, since E_k is proportional to the primary voltage. At no-load, I_2 is the magnetizing current. Fig. 60 shows the saturation curve expressed in terms of E_k and I_2 as well as the equivalent resistance line given by equation (34) above. It is evident that the voltage at which the machine operates is determined by the point of intersection of these two curves in exactly the same way as in a shunt-excited D.C. generator. If the resistance of the secondary circuit exceeds a certain value, self-excitation is not possible. It should be noted that similar self-excitation can occur when static capacitors are connected across the terminals of an induction motor.

The condition that self-excitation shall not take place is that the quadrature component of the injected voltage must not be sufficient to pass the full magnetizing current through the secondary resistance. In other words, the value of E_{kq} must be such that the motor, when running light, would operate at a lagging power factor when supplied in the ordinary way (see Fig. 17, p. 25). If a motor is highly saturated at normal voltage it may be possible for it to self-excite at a lower voltage even though the power factor is slightly lagging at normal voltage. In general, however, it is approximately true that a motor will or will not self-excite when disconnected from the supply, according to whether the power factor when running light is leading or lagging, and that the greater the leading current when running light under normal conditions of supply, the greater the voltage of self-excitation will be.

The frequency of self-excitation is determined by equation (35), since this gives the slip s in terms of the induced and injected voltages, and the actual speed of rotation is known. The frequency of self-excitation is approximately the same as that of the supply required to make the motor run at that speed under normal conditions.

Self-excitation on load

The above remarks apply to operation as a generator on open circuit. If an external load is connected, the conditions of self-excitation are affected by the impedance drops due to the load currents in primary and secondary, since these drops are added vectorially to the primary and secondary induced voltages. In general, a voltage is maintained of such a magnitude and frequency, that the machine, if connected to a system of that voltage and frequency, would pass the current, defined in terms of power and wattless current, which the load consumes. The voltage is approximately the same as that on open circuit, but, in general, the load produces a small drop both in voltage and frequency.

The use of an asynchronous generator as an independent source of alternating current has little practical importance, because the requirements can be met with a synchronous machine in a much simpler manner. Induction motors, both with and without commutator machines, are often used for duties where alternative motor and generator operation is required, but in such cases they are always connected to a supply of fixed voltage and frequency, and the value of current depends on the supply conditions. The property of self-excitation is seldom of practical value, but is much more often a source of trouble. Any motor which, as a result of an injected voltage, operates at unity or leading power factor when running light, tends to self-excite if the supply is disconnected. The voltage generally rises, and there is danger of burning out lamps or damaging insulation in any circuit which remains connected to the motor. This can, however, easily be prevented if the resistance of the secondary circuit is increased, or if the circuit supplying the frequency changer is broken before the main switch is opened. The rise in voltage is similar to that which would occur after switching off an over-excited synchronous machine, if the field were not disconnected at the same time. The phenomenon can occur in an induction motor with a frequency changer or a Scherbius machine, in a Schrage motor, or in a shunt motor.

21. Self-excitation in induction motors and commutator machines

It has already been mentioned that self-excitation can take place in the shunt commutator motor as a result of the connexion between the primary and secondary windings. The voltage builds up in the same way as in an induction motor with a frequency changer, and reaches a stable value which depends on the saturation curve of the motor. It is evident that this must be so, because both systems are equivalent to an induction motor with an injected voltage which depends on the primary voltage. The only difference is in the manner of interconnecting the primary and secondary windings. While the voltage induced in each actual conductor of the secondary winding has the same frequency, namely the slip frequency, in both types of machine, in the shunt motor it is converted by the commutator to supply frequency in the external secondary circuit. In the induction motor with a frequency changer, the conversion of frequency takes place in the frequency changer. In either system, the condition for self-excitation is that the currents flowing in the windings of the main machine shall set up a flux which induces voltages of such magnitude, phase and frequency as to maintain a state of equilibrium.

Self-excitation may occur in any A.C. commutator machine in a similar way, whenever a commutator armature winding and a stator winding are connected together to form a closed circuit, provided the condition stated above is satisfied. In general there is a certain limiting brush position, and a certain limiting value of resistance at which self-excitation can begin in any given machine. The action differs from that in a D.C. shunt generator because the voltage, current and flux are vectorial and not scalar quantities, and because there is an additional factor introduced, namely the frequency of the voltage and the current. In a condition of equilibrium the voltages, currents and frequencies must attain values such that the vector diagram of voltages in each circuit is closed.

Under the conditions described above, the current in the primary winding is small if no external load is connected, and the flux is due mainly to the secondary current particularly if the slip is small. The relative magnitudes of the primary and secondary ampere-turns depend on the transformation ratio of the machine, and on that of the transformer or regulator where one is included in the circuit. In the more general case, where the primary current is not small, the flux is produced by the resultant M.M.F. of the armature and stator windings, and the resultant voltage available for circulating the secondary current depends on the voltages induced by this flux in the two windings. If at any frequency the resultant voltage can pass a current of sufficient magnitude and of the correct phase to maintain the flux, self-excitation will take place.

An effect of this kind, known as series self-excitation, takes place in a polyphase series motor if the brushes are moved back beyond the neutral position while the machine is still running in a forward direction. It can also occur in a Scherbius machine if the brushes are set too far backward, as mentioned on p. 51. In both these examples the current rises to an excessive value, and the condition is one to be avoided.

Self-excited commutator generator

The phenomenon of self-excitation, which is normally undesirable, is brought into practical use, in certain applications of Scherbius machines with shunt-excitation. Such machines are used as shunt phase advancers, as speed regulating machines, or as independent low frequency generators.

The connexions of a shunt-excited low frequency three-phase generator of the Scherbius type, operating on open circuit, are shown diagrammatically in Fig. 61. The compensating winding is connected between the terminals and the brushes, while the

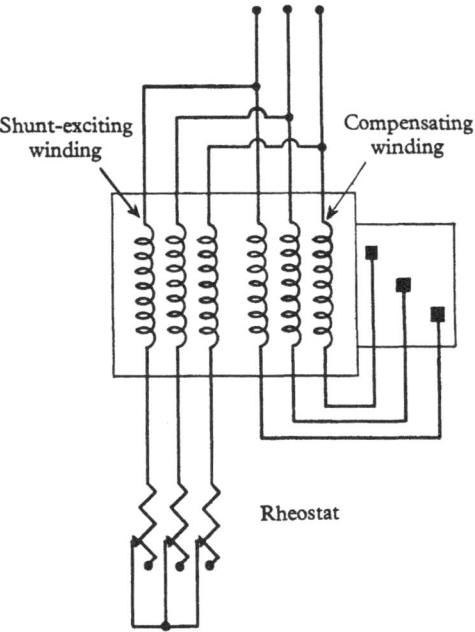

Fig. 61. Connexions of Scherbius low frequency
shunt-excited generator

exciting winding, the axis of which is displaced from that of the compensating winding, is shunt-connected in star across the terminals with a resistance in series. If the exciting winding is located so that the armature voltage is in phase with the exciting current, the relative position of the windings is the same as in a D.C. machine and self-excitation occurs with direct current, so as to build up D.C. voltages at the terminals, if the resistance of the exciting circuit is small enough.

Now suppose that, by changing the location of the exciting winding, the conditions are changed so that, with a given phase sequence of the currents, the exciting current lags behind the induced voltage in the main circuit by a phase angle α. Self-excitation is now only possible if the current alternates at a frequency such that the terminal voltage passes an exciting current lagging behind the main induced voltage by the angle α. On no-load, when the terminal voltage equals the induced voltage, α is the impedance angle of the exciting circuit at the frequency of operation.

Fig. 62 is a vector diagram of voltages in the exciting circuit for the no-load condition. OB represents E, the main induced voltage,

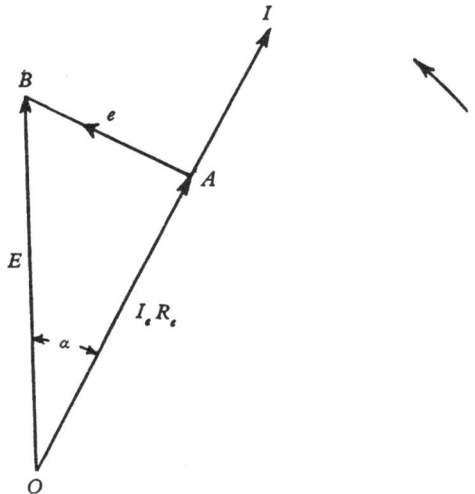

Fig. 62. Voltages in the exciting circuit of self-excited generator on no-load

OA the resistance drop $I_e R_e$, and AB the induced voltage e in the exciting winding due to both main and leakage fluxes. At a constant speed of rotation E depends only on the flux, but e depends on the frequency as well as the flux. Hence $e = KfE$, where K is a constant for a given speed depending on the numbers of turns in the armature and exciting windings and including a leakage factor.

Hence
$$E \cos \alpha = I_e R_e, \tag{36}$$

$$E \sin \alpha = KfE, \tag{37}$$

and
$$f = \frac{\sin \alpha}{K}. \tag{38}$$

Thus the frequency is fixed by the design of the machine and is independent of the resistance of the field circuit. Equation (36), in conjunction with the saturation curve of the machine, determines the voltage E, as shown in Fig. 63. The resistance line corresponds to a resistance $R_e/\cos\alpha$, and the condition of operation is determined by the intersection of the two curves. The voltage can be varied within limits by varying the resistance in the field circuit.

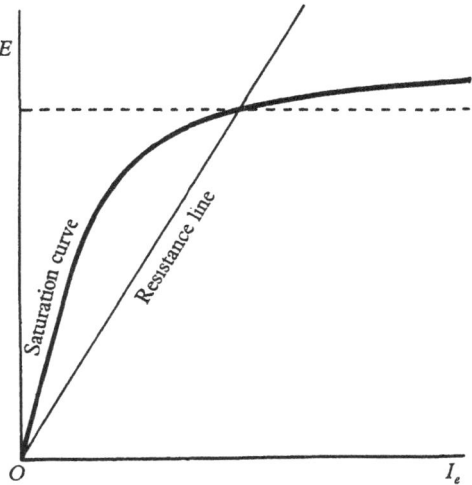

Fig. 63. Self-excitation voltage of Scherbius low frequency generator

When an external load is connected, both the magnitude of the voltage and the frequency fall off to some extent compared with the no-load values. If necessary, compounding by means of a series winding, suitably located, can be used to neutralize these drops. This type of machine is useful for providing a low frequency polyphase supply of substantially constant voltage and frequency.

22. Synchronous operation of asynchronous machines

If the polyphase rotor winding of an induction motor is supplied with direct current, stationary north and south poles are set up on the rotor; the position of the poles depends on the particular phases which are supplied. If, at the same time, the stator is connected to a polyphase A.C. supply, the motor runs at exactly synchronous speed, provided that the torque demanded does not exceed a certain value. This is the principle used in the synchronous induction motor, which starts up as an induction motor, but is synchronized when up to speed and runs at unity or leading power factor as a synchronous motor. As the load increases the speed does not change, except that the angular position of the rotor moves backward by a small angle.

Asynchronous operation at synchronous speed

An induction motor with a commutator machine in the rotor circuit can be made to run at exactly synchronous speed, provided that the voltage injected into the secondary circuit has a suitable value. At this speed, the frequency in the secondary circuit is zero, and each phase carries direct current. Normally, however, the machine is not 'synchronized'—that is, forced to run at synchronous speed—but operates asynchronously, in the sense that a small change in load would pull the motor away from synchronism to a speed slightly different. This condition exists in a motor regulated by means of a Scherbius machine or a frequency changer, as well as in a Schrage motor, when they are operating at synchronous speed. Provided that the phases are balanced, and the connexions are such that the injected voltage is always of the same frequency and phase sequence as the induced voltage, as it is under ideal conditions in the machines mentioned, the transition from a speed slightly above synchronous speed, through synchronous speed, to a speed slightly below is perfectly continuous. During the transition, the secondary frequency changes continuously from a value corresponding to the super-synchronous speed, down to zero frequency, and then to a value of frequency corresponding to the sub-synchronous speed. As the speed changes there is a change of the phase sequence of the secondary polyphase voltages.

During the transition through synchronism, the zero value of frequency should be considered as a limiting condition of alternating current rather than as direct current, and in this sense the term 'zero frequency alternating current' can be used. The three direct currents in the three phases can be considered as instantaneous values of a three-phase system which alternates so slowly that the instant chosen is prolonged indefinitely. The three currents may be distributed between the phases in any ratio, and hence the phase relationships between the alternating quantities persist at zero frequency, the equivalent phase of any quantity depending on the distribution of the phase values at the instant considered in the indefinitely slow cycle. In this conception, the vector diagrams of voltages and currents still hold good at zero frequency as at every other frequency. As already stated, the induction motor does not differentiate between synchronous speed and any other speed and operates at or near synchronous speed with asynchronous characteristics and without any discontinuity at synchronous speed.

Synchronization at synchronous speed

Under certain conditions, on the other hand, the motor may lock into synchronous speed, and remain synchronized over a range of loads in the same way as a synchronous motor. This occurs, for example, if the phase sequence of the voltages obtained from the commutator machine connected in the secondary circuit is opposite to that of the voltages induced in the secondary winding of the motor. It is evident that the only frequency at which the voltages can balance is zero, and the motor therefore locks into synchronous speed. The injected voltage serves to pass direct currents through the secondary phases of the motor and no speed variation occurs when the voltage is varied. The voltage applied to the secondary winding acts as an exciting voltage, and passes an exciting current which is independent of the load. As in any synchronous motor, there is a definite maximum torque above which the motor pulls out of synchronism, depending on the value of exciting current. This principle has been used in a form of self-contained self-excited synchronized motor obtained by reversing two secondary leads of an Osnos motor (see Section 42).

A weaker synchronizing action takes place if the injected voltage is of correct phase sequence, but is unbalanced between the phases, since an unsymmetrical system of voltages can be resolved into two components, one of forward phase sequence and one of backward phase sequence. Owing to the backward component, the machine synchronizes over a range of loads, the extent of which depends on the extent of the dissymmetry.

Dissymmetry of the secondary voltages may be due to faulty connexions, unbalanced windings or resistances, or most commonly, to inaccurate spacing of brushes. The phenomenon of synchronization provides a useful indication as to whether any wrong connexions exist in the secondary circuit. In variable-speed motors which are intended to operate with asynchronous characteristics, even a weak synchronizing action is undesirable, since it disturbs the continuity of the speed variation, and—what is often more important—causes the speed and supply current to pulsate when the speed control device is set for a speed near synchronism. The phenomenon of 'synchronous hunting' is discussed in the next section.

Synchronization away from synchronous speed

Under the conditions just described, the motor is forced to run at synchronous speed as the result of direct currents flowing in the rotor winding, that is, it is synchronized with the supply frequency. If a voltage of any frequency other than zero is impressed on the rotor, the motor may be forced to run synchronously at a speed away from synchronism. A motor supplied in this way is known as a 'doubly fed' motor.

This principle makes possible the synchronization of two or more induction motors in such a way that the motors always run at exactly the same speed as each other in spite of variations of load. A synchronizing action of this kind occurs if the rotor windings of several induction motors are simply connected together, or if they are connected to a common resistance or reactance. In these cases the synchronizing voltage for any one motor is the slip-frequency voltage of the other motors, or the drop in the external impedance. By using a commutator machine instead of an impedance to provide the synchronizing voltage, a wider range of action or greater flexibility of control is possible.

23. Hunting phenomena

Hunting in A.C. machines is characterized by a pulsation of the primary current above and below a mean value at a frequency different from, and generally much lower than, the fundamental supply frequency. In synchronous machines hunting is accompanied by a phase swinging of the rotor about the position it would occupy if the rotation were perfectly uniform. In asynchronous machines pulsation of the primary current is usually associated with a periodic variation of the speed.

If the supply voltage to an induction motor remains constant the flux must be a purely rotating one in order that a constant voltage may be induced, even though the current pulsates; hence the magnetizing current must also remain constant. Since the secondary ampere-turns at any instant equal the difference of primary and magnetizing ampere-turns, the pulsations of primary current must be accompanied by pulsations of secondary current.

Pulsation of both primary and secondary currents can be caused by a variation of load, but this condition cannot be described as hunting. Under these conditions there is also a small periodic variation of speed corresponding to the relation between torque and slip. True hunting is caused by effects inside the machines even when the external load is constant. When hunting occurs, the pulsations of current are accompanied by pulsations of internal torque and hence variations of speed. The fundamental of the pulsation, which appears as a modulation of the primary frequency (see Fig. 64) can be split up into two sinusoidal components of different frequency as given by the equation

$$A \cos \omega t + B \cos \omega' t = (A - B) \cos \omega t + 2B \cos \frac{\omega - \omega'}{2} t \cos \frac{\omega + \omega'}{2} t.$$

$$(39)$$

If the speed of the rotor is assumed to be uniform, each of these components of different frequency is balanced by a current in the secondary of a definite frequency. Hunting means, in effect, the introduction in the primary circuit of a current of non-fundamental frequency, and it is thus due to the presence in the secondary of a current of a frequency different from the normal slip frequency. The conditions are somewhat modified if the speed pulsates, as it generally does, but the general conclusion still holds. The disturbing current may be introduced by either of two causes:

(1) Dissymmetry of the secondary circuits.

(2) Independent self-excitation.

Hunting due to dissymmetry of the secondary circuit

If an induction motor operates with secondary resistances which are unbalanced between the phases, the secondary voltages induced by the rotating flux are all equal, but, because of the unequal resistances, the currents are unbalanced. As mentioned on p. 116, the unsymmetrical system of currents is equivalent to two sym-

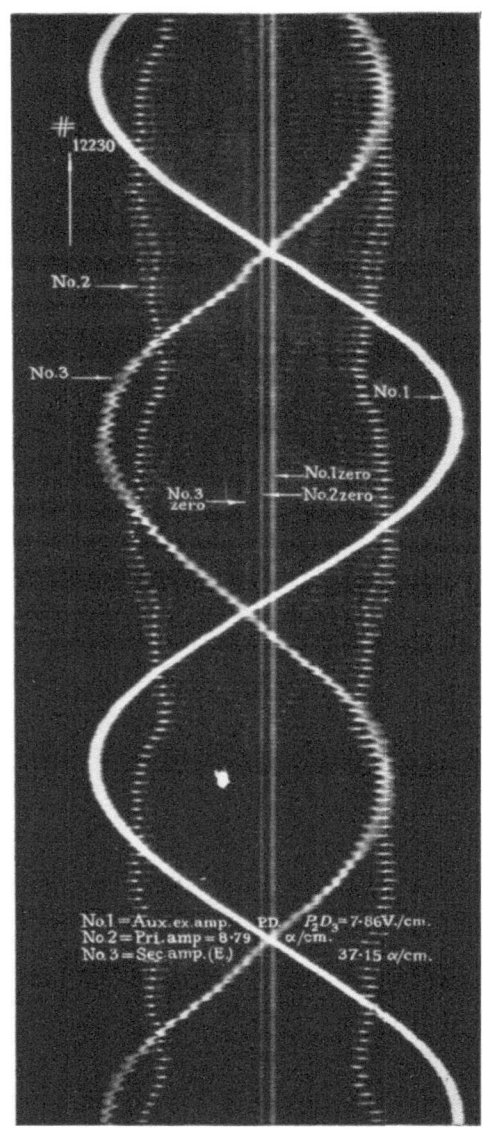

Fig. 64   *to face p.* 119

metrical systems of opposite phase sequence, but of the same frequency. The forward sequence component is balanced against a primary current of fundamental frequency, but the backward sequence component, which would set up a flux rotating relative to the rotor in the opposite direction, is balanced by a primary current of non-fundamental frequency.

If the rotor continues to run, in spite of the current pulsation, with a constant slip s, the main field rotates forward relative to the rotor with speed sn_0, where n_0 is the synchronous speed. The field corresponding to the backward sequence component of secondary current would rotate backwards relative to the rotor with speed sn_0, or at an actual speed relative to the stator $n_0(1 - 2s)$. Hence in equation (39), $\omega' = \omega(1 - 2s)$ and the total primary current becomes

$$A\cos \omega t + B\cos \omega(1 - 2s)\,t = (A - B)\cos \omega t + 2B\cos s\omega t \cos \omega(1 - s)\,t,$$
(40)

giving a curve similar to that shown in Fig. 64. An A.C. ammeter which gives a steady reading for a current at fundamental frequency pulsates at twice slip frequency when this type of hunting occurs. Pulsation of the primary current at twice slip frequency is in fact characteristic of hunting due to dissymmetry of the secondary circuit.

As already stated, the speed does not usually remain constant, because of the variation of torque due to the pulsations of current. The variations of speed keep time with the pulsations of primary current. The whole action is still periodic, but the variations are no longer sinusoidal, because the rate of change becomes slower when the slip is reduced and quicker when the slip is increased. Fig. 64 shows the oscillations of primary current at twice the frequency of the secondary current.

Any dissymmetry in the secondary circuit, of the kind mentioned on p. 116 in connexion with synchronization, may cause hunting. It may occur in an induction motor with any type of regulating gear in the secondary circuit, or in the Schrage motor. It is particularly liable to occur at speeds near synchronism because the secondary voltages are then low and errors assume larger proportions. Care in the grading of resistances, and in the spacing of brushes and commutator segments, usually suffices to reduce the pulsation to a harmless value.

Synchronous hunting

The effect of dissymmetry of the secondary circuit is most pronounced when the machine operates near to synchronous speed. The pulsations of current and speed are then called *synchronous hunting*. Over a certain range of adjustment of the control, the machine locks into synchronism, but just outside this range, pulsations occur as though the machine were continually pulling out of synchronism and slipping to the next pole.

When investigating the phenomenon of synchronous hunting with a stroboscope, it is found that the points of the stroboscopic disk, which remain stationary when the machine is synchronized, notch round from one pole to the next if the control is adjusted so as to vary the speed away from synchronism. The action is one of intermittent synchronization, which, however, is not strong enough to prevent the rotor from pulling out after a time and moving on, to synchronize again for a similar period a pole pitch away. Even when the motor is virtually synchronized, however, the point of the stroboscope is still moving slowly but is slowed down because the slip is reduced to a low value during that part of the cycle. If the control is readjusted so as to increase the slip, then the irregularity of the pulsation becomes less and approaches more to the condition discussed on p. 118 under the assumption that the speed was uniform. At the same time the magnitude of the pulsation usually decreases as the speed departs further from synchronism.

The relation between slip frequency and pulsation frequency can be clearly seen during synchronous hunting. A half period of the secondary cycle is the time taken for the rotor to move from one position to another a pole pitch away from where it would be rotating at synchronous speed, that is, the time for the points of the stroboscope to travel a pole pitch. After this period, conditions are as before, and the cycle of changes is then repeated with the same period indefinitely. Hence the period of the pulsations must be half the slip period, and the primary current and speed fluctuate at twice slip frequency.

Hunting due to independent self-excitation

In Section 20 it was explained how a machine or equipment can generate a voltage by self-excitation when it is not connected to any supply, and has its primary terminals open-circuited. If, instead, the primary terminals are short-circuited, the conditions are very much altered, but there is still the possibility of independent self-excitation under certain circumstances. The fundamental condition is, as before, that the voltages induced must be such as to pass currents in the various circuits which are of exactly the right magnitude and phase to maintain the voltages. The frequency of self-excitation depends on the impedances and other circuit constants in the equipment.

If the machine is connected to a supply system, the primary circuit is closed through the system, which for any frequency other than the supply frequency constitutes a short-circuit. Thus if the machine would self-excite when the primary is short-circuited, it will also self-excite when it is connected to a supply system, unless the frequency of self-excitation happens to coincide with the supply frequency. It is therefore possible for the machine to carry two independent alternating currents of different frequency, one the normal working current which produces the torque of the machine, and the other an additional current due to self-excitation which increases the heating of the machine without doing any useful work.

Hunting due to independent self-excitation has been known to occur in induction motors with Scherbius machines, particularly when these are used as phase advancers, and in Schrage motors. It is not often a serious difficulty, as it is usually possible to operate under conditions where self-excitation of this kind cannot take place.

Chapter 6

POLYPHASE SHUNT AND SERIES MOTORS

24. General arrangement

The motor formed by combining a polyphase stator winding with a rotor carrying a commutator winding, can be used with the stator and rotor either in shunt or series connexion as explained in Section 10. The motor is essentially the same with either connexion, but the control arrangements differ and different characteristics are obtained. The connexion between stator and rotor must be made through a transformer or regulator, because the commutator voltage is low compared with the stator voltage. In the shunt motor, the brush gear is generally fixed and the control is obtained by means of a variable ratio transformer or regulator. The series motor, on the other hand, operates with a fixed ratio transformer, and is controlled by moving the brushes. The two types of motor must be treated separately, both for describing the operation and for developing a theory.

Polyphase shunt motor

The shunt motor is equivalent to an induction motor, except that the frequency of the voltage and current at the brushes is converted to the constant supply frequency by the action of the commutator. With a given setting of the regulating apparatus the injected voltage applied to the secondary terminals is constant, apart from impedance drops. Thus the shunt motor can be treated as an induction motor with a constant injected voltage. The motor is controlled by adjusting the injected voltage.

Modern motors of this type are almost always controlled by means of an induction regulator, which, either by itself, or in combination with other apparatus, provides a variable voltage of suitable magnitude and phase for controlling the motor speed, while at the same time maintaining a satisfactory power factor. The regulator is generally a double unit, composed of two induction regulators mounted on the same shaft. A typical diagram of connexions of the equipment is given in Fig. 65. The two primary windings of the regulator are connected in parallel to the supply and the two secondary windings are connected in series. The transformer shown in this diagram is used to provide a small constant component of voltage in quadrature with the variable voltage of the regulator.

The vector diagrams of the shunt motor are identical with those given in Figs. 13–15 for the induction motor with a constant injected voltage, notwithstanding two important practical differences. The first difference is that the secondary frequency is

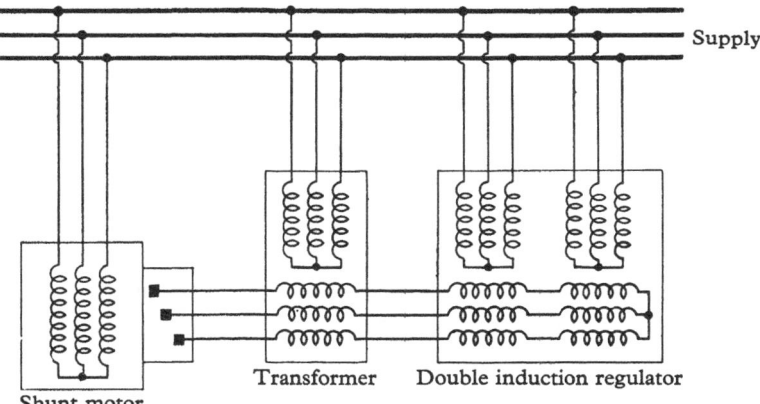

Fig. 65. Connexions of shunt motor equipment

always the same as the supply frequency. Hence there is an actual physical phase relationship between primary and secondary voltages, whereas in the induction motor, as explained on pp. 4 and 22, the phase angle is merely the angle between vectors representing voltages of different frequencies. The second difference is that the secondary reactance is not proportional to the slip but contains a constant component, which has an important effect on the characteristics, as discussed in Section 26.

In accordance with the treatment in other machines, the injected voltage E_k in the vector diagram is not the actual output voltage of the regulator, but is the referred value, since the motor to which the diagram applies is the equivalent unity ratio motor. In what follows, it is assumed that this transformation has been made, that is, that E_k is obtained by multiplying the actual voltage by the transformation ratio of the motor.

Fig. 66 is a vector diagram showing the voltages of a double induction regulator. The primary voltage is V shown vertically upwards. Assume that in one position of the regulator the secondary voltages OX and YO are equal to each other and in phase with the primary voltage V, and that they add. Then YX is the maximum secondary voltage obtainable from the double

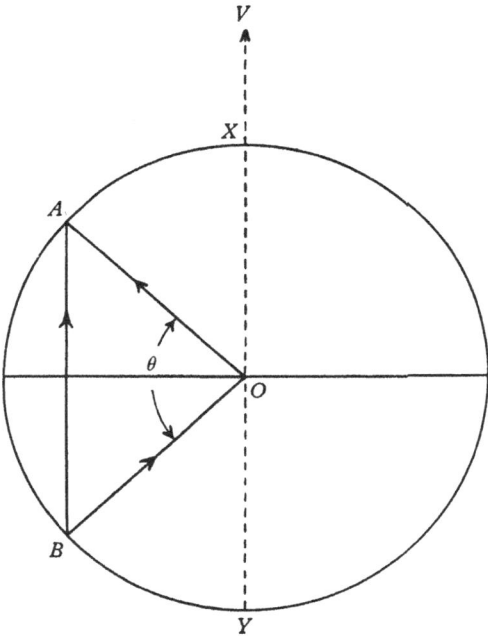

Fig. 66. Vector diagram of double induction regulator voltages

regulator. As the rotors rotate the two voltages remain constant in magnitude but vary in phase according to the position of the rotor.

The two rotors are mounted on a common shaft, and the windings are arranged so that the fluxes of the two machines rotate in opposite directions. A movement of the rotor therefore advances the phase of one voltage and retards that of the other by the same angle. Thus the vector OX rotates to OA and YO to BO; it follows that the resultant voltage BA is reduced in magnitude but remains in phase with V.

When the regulator is connected so as to supply a voltage to the secondary of a shunt commutator motor, the phase of the effective injected voltage E_k in relation to the motor primary voltage V, as expressed in the motor vector diagrams, depends not only on the actual angle between regulator primary and secondary voltages, but also on the position of the motor brush gear. There is one brush position, known as the *neutral brush position*, in which

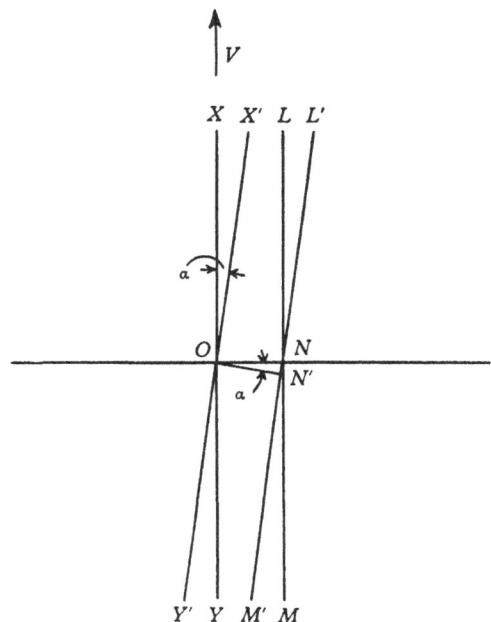

Fig. 67. Locus curves for E_k in shunt motor

the injected voltage E_k is in phase with V and in such a direction as to reduce the speed below synchronism. In this position the same characteristics are obtained with either direction of rotation of the motor. With the brushes in the neutral position, E_k, which may be represented by BA in Fig. 66, remains in phase with V and varies in magnitude between a positive maximum and a negative maximum as the double regulator is rotated. The locus of E_k is therefore the line XOY in Fig. 67. When the brushes are displaced from neutral by the electrical angle α the locus of E_k becomes $X'OY'$.

It is not necessary for the secondary voltage of the regulator to be in phase with the primary voltage, because a phase angle between them can be compensated by a corresponding change in the neutral brush position. With the modified neutral position, the effective injected voltage to be used in the equivalent circuit is still in phase with the motor primary voltage, although the actual voltages are no longer in phase.

A motor used with a double regulator, as just described, operates at a low power factor, and it is generally worth while, except in small motors, to provide additional means, such as the transformer shown in Fig. 65, in order to improve the power factor. The transformer introduces a component of voltage in quadrature with that provided by the regulator, and this is added to the regulator voltage. The contribution to E_k of the transformer voltage is represented in Fig. 67 by the constant vector ON at right angles to OX, and the locus of E_k now becomes LNM, when the brushes are on neutral. With brush shift α, N moves to N', where angle $NON' = \alpha$, and the locus becomes $L'N'M'$.

The neutral position must be defined in relation to the regulator voltage, which always has the same phase, and not to the resultant injected voltage, because the phase of the latter is variable.

The arrangement described here makes it possible to obtain any desired straight line locus for the injected voltage. The two end-points of the locus can be chosen to give to the motor any desired values of speed and power factor at the two ends of the range. When the regulator is rotated through 180 electrical degrees between the two limiting positions, the speed of the motor changes smoothly from a minimum speed below synchronism to a maximum speed above synchronism.

Practical arrangements

Many variations are possible in the practical apparatus used to obtain the regulating voltage required by the shunt motor. The combination shown in Fig. 65, involving three items, that is, two regulators and a transformer, is the most straightforward, and is frequently used, particularly for large high voltage motors. For smaller powers, many attempts have been made to simplify the regulating apparatus.

Auxiliary windings. The transformer can be replaced by an auxiliary winding either on the motor or the regulator. The auxiliary winding, when used, is located in the same slots as the primary winding of the motor or of one regulator, and acts as the secondary of a transformer. This arrangement eliminates one piece of apparatus and is commonly used.

Combined motor and regulator unit. The separate auxiliary voltage would not be necessary if the two main components of regulating voltage (BO and OA in Fig. 66) were of unequal magnitude and rotated at different speeds. It is not very convenient to carry this into practice when using two regulators, but the principle has been used in a type of machine made in considerable numbers in Germany. In this machine, one component of regulating voltage is provided by an auxiliary stator winding on the motor, and the effective rotation of the voltage vector for this component is brought about by rotating the motor brush gear. The second component of regulating voltage is obtained from an induction regulator mounted above the motor in an extension of the motor frame. The operating mechanism rotates the brush gear and regulator together, the two elements being coupled by means of a chain or strap. With this arrangement, the vectors representing the two components of regulating voltage can be unequal and can rotate at unequal rates as the speed control mechanism is operated. Hence a good power factor can be obtained at all speeds without the use of any further auxiliary voltage.

Single unit regulator. There are several patented schemes of connexion which permit the use of a single unit regulator, sometimes by itself and sometimes in conjunction with an auxiliary winding on the motor. An appropriate law of variation of the voltage is obtained by combining together two or more windings on the stator and rotor of the regulator.

Although it is an advantage to eliminate one regulator, there is added complication in the remaining regulator, and as a result, the arrangement has not had any extensive application except in small sizes.

Polyphase series motor

The only additional apparatus needed with a series motor is the transformer used for interconnecting the primary and secondary windings, as indicated in Fig. 34. This transformer is usually quite small and is often mounted inside the motor frame. In spite of its greater simplicity compared with the shunt motor, however, the application of the series motor is limited by the fact that its characteristics are of the series type.

The secondary current of the motor is proportional to the primary current under all conditions, if the magnetizing current of the transformer is neglected, and it is in phase with, or at a constant phase angle to, the primary current depending on the transformer connexion. The effective phase angle between the referred secondary current and the primary current, depends not only on the transformer but also on the position of the motor brushes. There is always one position, known as the *neutral brush position*, where the ampere-turns due to primary and secondary currents are in direct phase opposition. With the brushes in the neutral position a large current flows when the supply is connected, and there is no torque, but when the brushes are displaced from neutral a torque is exerted, and the characteristic of the motor depends on the brush position. Control of the series motor is obtained by moving the brushes between positions corresponding to small and large angles of shift. A displacement from neutral in the opposite direction causes the motor to run with reversed rotation. When the field rotates in a direction opposite to that of the armature, however, the characteristics are poor, and this method of operation is not normally used.

Thus a series motor normally has a single set of movable brush gear. It is sometimes, however, constructed with two sets of brush gear, one fixed and one movable, the transformer secondary winding being connected between them. This arrangement has the effect of changing the transformation ratio of the motor as well as the angle of shift, and results in improved characteristics at low speeds. A similar effect can be obtained in steps with a motor having a single set of brush gear, by changing taps on the transformer.

Special features to assist commutation

It is an advantage in large shunt or series motors to increase the number of secondary phases in order to assist commutation. The commonest number is six, because a six-phase arrangement of brush gear is equivalent to an open three-phase connexion, and can be used in conjunction with a three-phase regulator or transformer. By using six or more phases it is often possible to increase the number of brush arms and hence to increase the output obtainable from a given size of commutator.

In large shunt or series motors, it becomes almost essential to use some kind of damping winding in order to assist commutation. The many types of damping winding used are discussed in Section 17.

General operation

Normally, shunt and series motors require no special starting gear, apart from the ordinary three-phase supply switch. The control mechanism, whether a regulator or moving brush gear, may be operated either by hand or by a small pilot motor, and it is usual, except for small motors, to ensure by appropriate interlocking that it is set in the low speed position before starting by closing the switch, in order to limit the current taken at standstill. The transformer voltage of commutation is greater when switching on than during normal running, and causes some sparking during the starting period, but this is generally permissible because of its short duration.

In special equipments, particularly those with short speed ranges, it may be necessary to use additional starting apparatus, either auto-transformers or resistances similar to those used with induction motors.

25. Theory of the shunt motor

The theory of the shunt motor is essentially that of the induction motor with a constant injected voltage. The principal difference is that the total secondary leakage reactance, which must include the leakage reactance of the regulating apparatus, is not proportional to the slip, but has in addition to the varying reactance sX_2, a constant component X_3. The reactance X_3 is important in determining the practical characteristics of the shunt motor, and provides the main reason why they are inferior to those obtained with the Schrage motor.

Secondary reactance

Apart from the effect of the regulator, the secondary leakage reactance of the shunt motor itself differs from that of the induction motor in that it is not directly proportional to the slip. A typical curve of variation is given in Fig. 68. At synchronous speed, the reactance is not zero, but has a value X_{3m}—generally about a third of X_2. The total motor secondary reactance at slip s is $sX_2 + X_{3m}$. The leakage reactance X_{3r} of the regulating apparatus is also independent of the motor slip; it is combined with X_{3m} to give the constant component X_3 of the total secondary reactance $(sX_2 + X_3)$. In practice, X_{3r} constitutes the major part of X_3.

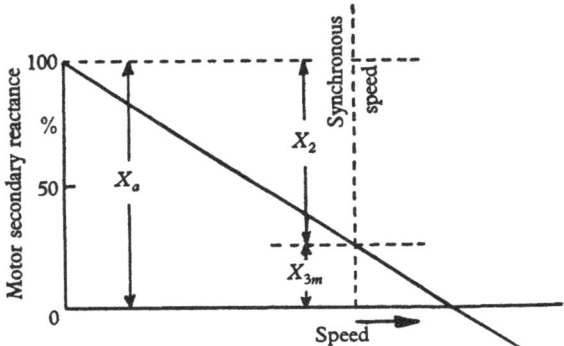

Fig. 68. Variation of secondary reactance with speed in shunt motor

Vector diagrams

The theory of the shunt motor differs from that given in Section 4 for the induction motor with an injected voltage mainly in that the secondary reactance drop sX_2I_2 is replaced by $(sX_2 + X_3)I_2$. Apart from this Figs. 13–15 are directly applicable to the shunt motor.

Equivalent circuit

Except for the difference in the secondary reactance, the equivalent circuit of Fig. 16 also applies directly to the shunt motor, in so far as the determination of the motor currents is concerned. It is, however, necessary to calculate the total current taken from the supply, and, in order to do this, the diagram of Fig. 16 is extended to include the source of E_k, as shown in Fig. 69. The combination of regulator and either transformer or auxiliary winding, which provides the resultant injected voltage, is treated as a single unit represented by a separate resistance and leakage reactance, a magnetizing circuit, and an ideal transformer having a transformation ratio k_r. The circuit used to represent the regulating apparatus is an approximate one similar to the simplified induction motor circuit of Fig. 7. In the diagram, the resistance is included with the motor secondary resistance and the brush resistance to give a total value R_2.

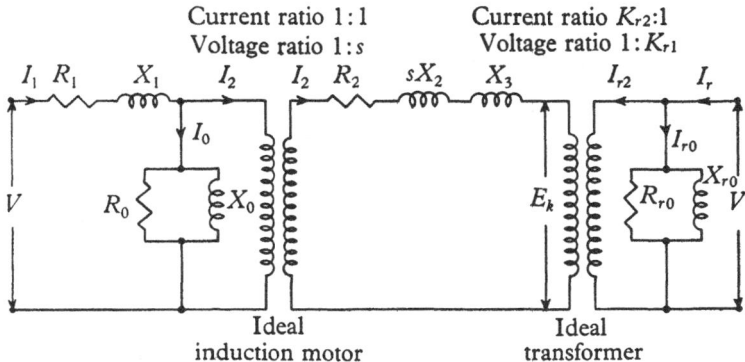

Fig. 69. Equivalent circuit of shunt motor equipment

The transformation ratio of the ideal transformer in Fig. 69 is a vector ratio because there is not only a scalar ratio k_r between V and E_k, but also a phase displacement ρ.

The ratio

$$\frac{E_k}{V} = k_r e^{j\rho} = K_{r1}, \tag{41}$$

while

$$\frac{I_{r2}}{I_2} = k_r e^{-j\rho} = K_{r2}. \tag{42}$$

The angle ρ depends not only on the connexions of the regulating apparatus but also on those of the motor and on the brush position. I_{r2} is the primary current of the ideal transformer, and the regulator current is I_r, obtained by adding to I_{r2} the magnetizing current I_{r0}. Hence the total current taken from the supply is

$$I_t = I_1 + I_{r2} + I_{r0}. \tag{43}$$

Fig. 70 is a vector diagram of the currents in a shunt motor and its regulator for operation as a motor below synchronism. I_{r2} opposes I_2, but differs from it in phase by the small angle ρ. The sum of I_1, I_{r2} and I_{r0} gives the resultant current I_t. For motoring operation above synchronism E_k would be reversed relative to V, and ρ would be nearly 180°; consequently I_1 and I_{r2} would then add together almost arithmetically. Hence a motor operating at constant torque takes more power from the supply at high speed than at low speed. It is obvious that this must be so, because the horse-power varies in proportion to the speed.

Equations of the shunt motor

The secondary impedance is

$$Z_2 = R_2 + j(sX_2 + X_3)$$

in which $X_3 = X_{3m} + X_{3r}$,

$$I_2 = \frac{V(s - cK_{r1})}{sZ_1 + cZ_2} \qquad (44)$$

The total current from the supply is

$$I_t = I_1 + I_{r0} - K_{r2}I_2$$
$$= \frac{VY_0}{c} + I_{r0} + I_2\left[\frac{1}{c} - K_{r2}\right]. \qquad (45)$$

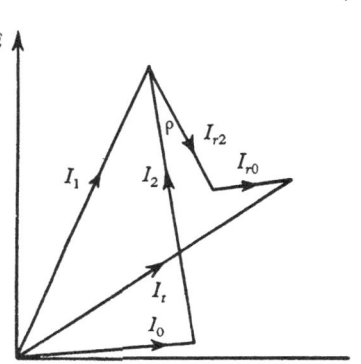

Fig. 70. Vector diagram of shunt motor and regulator currents

These equations, together with those given in Section 4, are sufficient for calculating the complete characteristics of the shunt motor.

26. Characteristics of the shunt motor

The characteristics of the shunt motor are similar to those shown in Figs. 21, 22 and 23, for an induction motor with a constant injected voltage. There are two principal factors which make them different:

(1) The constant component X_3 of secondary reactance causes an impairment of the power factor throughout the speed range, when the motor is loaded.

(2) The current I_r returned to, or taken from, the supply by the regulator must be added to the motor primary current I_1 in order to obtain the total current I_t.

The regulator leakage reactance, which provides the greater part of X_3, is the most important limiting factor in the design of a shunt motor equipment. It must be kept to the lowest possible value, and even then, causes the characteristics to be definitely inferior to those of the Schrage motor, which corresponds more closely to an induction motor with constant injected voltage. The curves in Figs. 71, 72 and 73, which give the characteristics of a 350 H.P. motor, show clearly the effect of the regulator reactance. In small motors, where the resistance drops are relatively larger, the effect of the reactance is not so great.

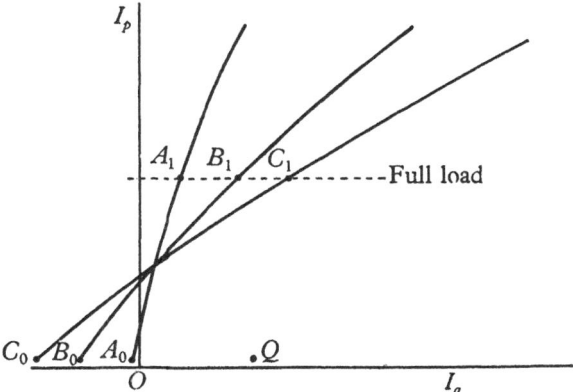

Fig. 71. Current locus curves of shunt motor

A given value of secondary current corresponds approximately to a definite torque irrespective of the speed. The shunt motor is thus essentially a 'constant torque motor', and in the following description it is assumed that the full-load at any speed corresponds to a given value of torque.

The control obtained by rotating the regulator of the shunt motor is very similar to that obtained by rotating the brush gear of the Schrage motor. Many of the points discussed on pp. 158–60 apply to the shunt motor.

Current locus curves

Fig. 71 shows typical curves of motor primary current in a shunt motor which is controlled by a double induction regulator and has an auxiliary winding on the motor. The vector locus of the injected voltage is a straight line similar to the line $L'N'M'$ of Fig. 67, the quadrature component E_{kq} being greater at low than at high speeds. Each point of the voltage diagram corresponds to a definite current locus curve, which, with the assumptions made, is a circle. It can be shown that, if the voltage locus is a straight line, all the circles intersect at a point. The three curves in Fig. 71 are those obtained at the two ends and the middle of the speed range.

In practice, the curves are not true circles, mainly because of the varying resistance of the brush contact, but the general shape is not greatly affected.

In the diagram the vector joining the origin to each point represents the motor primary current. The full-load points at top,

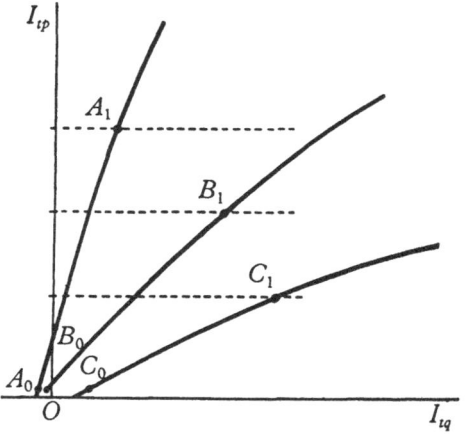

Fig. 72. Locus curves of total current in shunt motor and regulator

middle and bottom speeds are indicated by A_1, B_1 and C_1, and the no-load points by A_0, B_0 and C_0. The secondary current of the motor is represented approximately by the vector joining Q to each point, where OQ is the magnetizing current I_0 of the motor. At the top and middle speeds the power factor at full-load is corrected to a high value, but it is not permissible to correct the power factor so much at bottom speed, because the secondary current at no-load, represented by QC_0, would then be too great. It is usual to set the regulator and motor brush gear so that the secondary current at no-load does not greatly exceed the full-load value, unless the application is such that the motor is always loaded. It is clear that at low speeds the power factor of the motor varies over a wide range between no-load and full-load, because of the effect of the regulator reactance.

Fig. 72 shows the current locus curves for the total current taken from the line. They are obtained by adding the regulator current for each condition by the method explained in the preceding section. The full-load and no-load points are again indicated by A_1, B_1, C_1, A_0, B_0 and C_0. In Fig. 71 a definite input to the

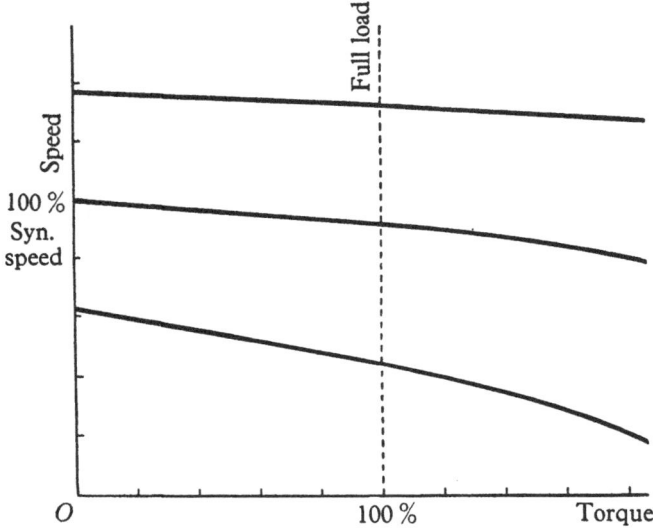

Fig. 73. Torque-speed curves of shunt motor

motor corresponds approximately to a definite torque, whatever the speed, and hence the full-load points all lie approximately on the same horizontal line. But in Fig. 72 the in-phase component of current corresponds to the total power supplied, and varies approximately in proportion to the speed.

The shunt motor can have a good power factor at high speeds, as shown in Fig. 72, but the power factor is bound to fall off at low speeds because of the combined effect of regulator reactance and magnetizing current. Owing to the drop in power output, however, the current taken by a motor operating at constant torque is appreciably less in magnitude at low speeds than at top speed.

Torque-speed curves

Fig. 73 shows typical torque speed curves at the top, middle, and bottom speed settings of a shunt motor. The motor has essentially a 'shunt characteristic' except that the drop in speed from no-load to full-load is a rather large proportion at low speeds. At high speeds the drop in speed from no-load to full-load corresponds to the secondary resistance drop, but is a larger percentage of synchronous speed than in an induction motor because of the additional secondary resistance due to the brushes and the regulator. At low speeds, the drop in speed is increased further because of the wide variation of the phase of the secondary current, which causes a varying component of secondary impedance drop in phase with the injected voltage. The poor speed regulation at low speeds is caused to a great extent by the presence of the regulator reactance. In a shunt motor operating at constant torque with a speed range of 3:1, for example, the full-load speed at the bottom end of the range is generally about 30–40% below the no-load speed. The proportion is approximately the same in both large and small motors.

27. Theory of the series motor

The fundamental vector diagrams of the polyphase series motor are the same as those of the induction motor or of the shunt motor; the difference in the development of the theory lies in the fact that the voltages applied to primary and secondary are no longer constant: they are individually variable, and it is their vector sum which equals the constant supply voltage. On the other hand, the motor current is common to primary and secondary.

The equivalent circuit of the series motor is given in Fig. 74; it is similar to that given in Fig. 69 for the shunt motor, but has different values of primary voltages and currents. The motor primary voltage is V_1 instead of the line voltage and the transformer primary voltage is V_2. The supply voltage is

$$V = V_1 + V_2. \tag{46}$$

The current in the transformer is the same as the primary current I_1 of the motor. Because of this it is more convenient to refer the secondary impedances of the motor and the transformer to the primary side of the ideal transformer as shown in Fig. 74. The magnetizing current of the transformer, which is much smaller than that of an induction regulator, is neglected. As before K_{r1} and K_{r2} are the vector transformation ratios taking into account not only the ratio of magnitudes of the primary and secondary voltages of the ideal transformer, but also the phase angle ρ which, as in the shunt motor depends on the motor brush shift and on the connexions of the transformer. The equations of the motor can be derived from this equivalent circuit.

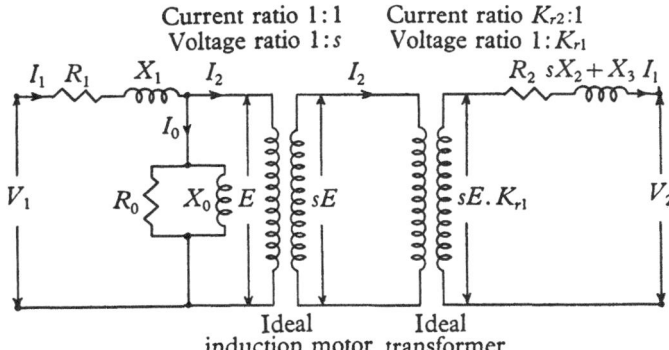

Fig. 74. Equivalent circuit for the series motor

Equations of the series motor

The total impedance traversed by the main current I_1 is

$$Z = Z_1 + Z_2 = R_1 + jX_1 + R_2 + j(sX_2 + X_3). \qquad (47)$$

Since
$$V_1 = E + I_1 Z_1, \qquad (48a)$$

$$V_2 = I_1 Z_2 - sEK_{r1}, \qquad (48b)$$

$$V = E\beta_1 + I_1 Z, \qquad (49)$$

where $\beta_1 = (1 - sK_{r1})$.

Also
$$I_1 = I_0 + I_2 = I_0 + K_{r2}I_1. \qquad (50)$$

Hence
$$E = \frac{I_0}{Y_0} = \frac{\beta_2 I_1}{Y_0}, \qquad (51)$$

where $\beta_2 = (1 - K_{r2})$.

Therefore, from (49) and (51),

$$I_1 = \frac{VY_0}{\beta_1\beta_2 + Y_0 Z}.$$ (52)

The torque is given by

$$W_T = mEI_2 \cos \phi_2,$$

where ϕ_2 is the angle between E and I_2. If R_0 is neglected,

$$W_T = mEI_1 \cos \phi,$$ (53)

where ϕ is the angle between E and I_1. For a given angle ρ, the angle ϕ, which depends on β_2/Y_0, is a constant.

28. Characteristics of the series motor

The current locus and torque-speed curves derived from equations (52) and (53) differ greatly from those of the shunt motor.

Current locus curves

The current locus curves, which can be calculated from equation (52) are circles passing through the origin as can be proved by equations (18) and (19). The origin corresponds to the condition of infinite slip. Four circles are shown in Fig. 75, for four different angles of brush shift, and on each curve, P indicates the synchronous speed point and Q the standstill point. It can be seen that the power factor is good at synchronous speed and above, but that it falls off at the lower end of the range.

Torque-speed curves

Since for a given brush position the torque depends only on the square of the current, the variation of torque is readily deduced from the equations. The four torque-speed curves corresponding to the current locus curves of Fig. 75 are given in Fig. 76. The curves are similar to those of a D.C. series motor except that the torque does not rise continuously down to standstill but has a maximum value at a positive speed.

For almost all industrial applications, the characteristics of the series motor are inferior to those of the shunt motor because of the large change of speed when the torque varies. The series motor is satisfactory for driving pumps or fans, where the torque required is a definite function of the speed, and increases rapidly with the speed. The series motor can also be used for other steady

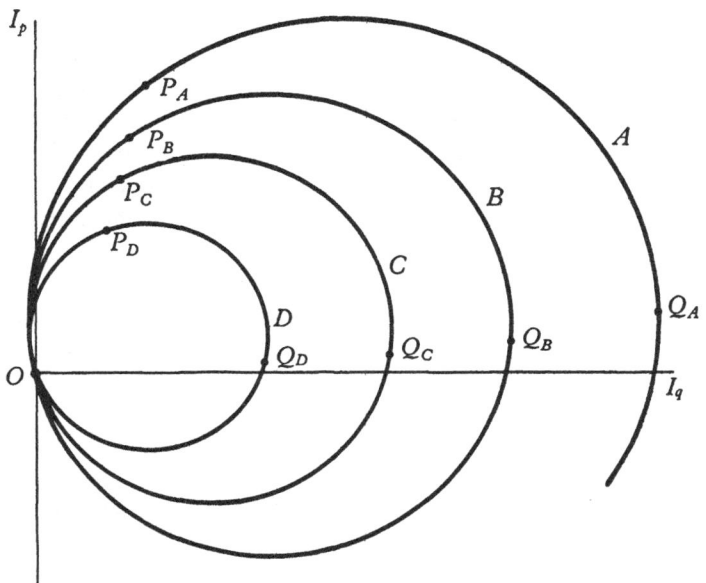

Fig. 75. Current locus curves of polyphase series motor

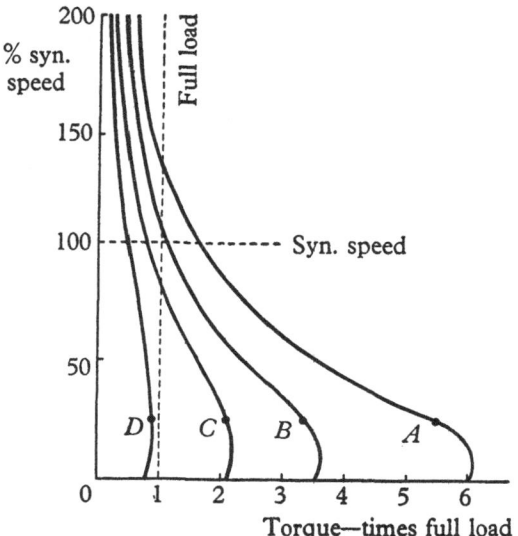

Fig. 76. Torque-speed curves of
polyphase series motor

drives, such as printing presses or textile machinery, where a constant torque is required over a range of speeds, provided that the operation takes place at a stable point of the torque-speed characteristic, that is, at a speed well above the value for maximum torque. The characteristics are considerably improved in this respect, and a longer stable speed range is obtained, if the motor is provided with double brush gear, or if taps on the transformer are changed, as described on p. 128.

The theory of the series motor given above is a good deal less accurate than that given for the shunt motor, because of the fact that the voltage applied to the stator winding is not constant. It is evident that the flux in the motor varies over a wide range as conditions change, and hence that saturation is a very important factor. A more accurate calculation can be made by using a trial and error method to allow for saturation.

The transformer magnetizing current also has been neglected. This does not cause serious discrepancy if the transformer is not saturated, that is, when the speed is within the normal working range of the motor. At large values of slip, however, there is considerable error due to the transformer magnetizing current. The effect is to reduce the torque at standstill and at low speeds, and to limit the high speed to which the motor rises at low loads.

Chapter 7

THE COMMUTATOR FREQUENCY CHANGER
AND THE SCHRAGE MOTOR

29. The commutator frequency changer

The applications of the commutator frequency changer depend on its ability to provide a low frequency polyphase voltage, and they can be classified under three main headings:

(1) Independent source of low frequency power.

(2) Source of the slip-frequency voltage (injected voltage) required for controlling the speed or power factor of an induction motor.

(3) Source of low frequency voltage for exciting a Scherbius machine.

The frequency changer has been used to a limited extent for supplying A.C. motors at a low frequency in order to obtain an abnormally low speed or for other special purposes. In these applications, the slip-rings are connected to a normal polyphase supply, and the rotor is driven at a speed near to the synchronous speed corresponding to the number of poles and the supply frequency. An alternative arrangement is to make the frequency changer self-driven by providing it with a stator winding. The application of the machine to this kind of duty has not been very extensive and it need not be described further.

When a frequency changer is used in conjunction with an induction motor, either to supply an injected voltage directly or to excite a Scherbius machine which supplies an injected voltage, the output frequency must always be the slip-frequency of the motor. This result is obtained, as explained in Section 11, if the frequency changer is mechanically coupled to the motor, has the same number of poles, and is connected so that the flux rotates backwards relative to the rotor. This arrangement is assumed in the description which follows.

Alternatively, the frequency changer may be driven from the induction motor through gears, provided that the ratio of frequency changer speed to motor speed is equal to the inverse ratio of the numbers of poles.

The normal frequency changer has no stator winding, and consequently cannot exert any torque. All the power taken from the commutator must therefore be supplied to the slip-rings or vice versa. Hence, if a frequency changer is used to reduce the speed of an induction motor, the power taken from the secondary of the motor is returned to the supply to which the frequency changer slip-rings are connected, less the losses in the machine.

Magnitude and phase of injected voltage

Consider first a frequency changer used for supplying an injected voltage directly, as illustrated in Fig. 35. Apart from the change of frequency, the frequency changer acts as a transformer with a constant ratio of transformation K_c. When a single winding is used and both slip-rings and commutator are arranged for three phases, the transformation ratio is unity, but when separate windings are used, K_c is the ratio of the effective numbers of turns. Thus, apart from impedance drops, the output voltage of the frequency changer is constant, if the voltage supplied to its slip-rings is constant.

It was explained on p. 4 that the primary vectors of the induction motor in Fig. 3 bear a definite phase angle relationship to the slip-frequency vectors in the secondary diagram of Fig. 4. The same relationship holds in the frequency changer, between the supply frequency voltage at the slip-rings and the slip-frequency voltage at the commutator. Hence, if the voltages supplied to the induction motor and the frequency changer come from the same source, the injected voltage from the frequency changer, when expressed on a vector diagram similar to Fig. 15, must be at a constant phase angle to the vector V. The frequency changer therefore provides an injected voltage which is constant both in magnitude and in phase. The referred value E_k of the injected voltage is obtained by multiplying the secondary voltage of the frequency changer by the transformation ratio of the induction motor.

The actual phase angle between V and E_k depends on the arrangement of the windings on the two machines, on the angular location of the mechanical coupling, and on the position of the brushes on the frequency changer. A transformer interposed between the two machines on the primary side may also affect the phase angle. Whatever the arrangement, however, there is always a brush position on the frequency changer for which E_k is in phase with V, and in such a direction as to reduce the speed of the motor. This position is known as the *neutral brush position*. Any other brush position can be defined by the displacement away from the neutral brush position measured in electrical degrees. The electrical angle, known as the *angle of shift*, is equal to the mechanical angle multiplied by the number of pairs of poles. In any position the phase angle between E_k and V is equal to the angle of shift.

In practice, the frequency changer, like the regulator described in Section 24, has resistance and reactance, and takes a magnetizing current from the supply. Its operation can therefore be taken approximately into account by a similar method to that used for the shunt motor.

The method may be summarized as follows:

(1) Include the impedance with the secondary impedance of the motor.

(2) Take an injected voltage $E_k = VK_c$, where K_c is a constant vector, determined by the transformation ratios of the two machines and the angle of shift.

(3) Add a constant current vector to the frequency changer input current, to represent its magnetizing current and core loss.

However, unlike the regulator in Section 24, the leakage reactance drop of the frequency changer to be deducted from the injected voltage on the secondary side, is mainly proportional to the slip, and the constant component X_3 is small (see p. 129). Hence the characteristics of an induction motor with a frequency changer are like those given in Figs. 21, 22 and 23. To determine the total current taken from the supply, the frequency changer current, including the magnetizing current, must be added vectorially to that of the induction motor.

When a frequency changer is used as an exciter for a Scherbius machine, the vector representing the voltage introduced by it into the exciting circuit is constant in magnitude and phase. This type of equipment is discussed in Chapter 8.

Control of frequency changer voltage

There are two methods of controlling the output voltage of a frequency changer so as to control the characteristics of the induction motor.

(1) *Controlling the voltage applied to the slip-rings.* The frequency changer then has a single fixed set of brush gear, and the slip-rings are supplied from a variable ratio transformer or induction regulator similar to that used with the shunt motor. It is necessary to control both the magnitude and phase of the applied voltage.

(2) *Control by brush movement.* The supply voltage is constant, and two movable sets of brush gear are provided. The output voltage, which is that between corresponding brushes on the two rings, can then be controlled completely by controlling the movement of the two brush rings.

Moving brush gear has not often been used on frequency changers which regulate an induction motor directly, but the arrangement is extremely important in connexion with the Schrage motor. The results obtained are therefore discussed in considerable detail in the next section. The application of the frequency changer with moving brush gear as an exciter for a Scherbius regulating machine is also important and is discussed in Chapter 8.

Compensated frequency changer

Large frequency changers sometimes have on the stator a compensating winding connected in series with the commutator brush gear, and carrying the main low frequency current. Interpoles and interpole windings may also be provided in order to improve commutation. Such windings can, of course, only be used if the brush gear is in a fixed position.

If the compensation is exact so that it exactly neutralizes the field due to the currents flowing through the commutator, the current supplied to the slip-rings is simply a magnetizing current and is considerably less than that supplied to an ordinary frequency changer. The compensated frequency changer exerts a torque and converts electrical power into mechanical power. Moreover, the output voltage is not quite independent of frequency, because the voltage induced in the compensating winding, which is variable, is added to or subtracted from the constant commutator voltage.

Notwithstanding these differences, the method of control of the compensated machine is the same as for an uncompensated one with fixed brush gear. Its advantages are that commutation is improved, and that the current taken from the supply is reduced.

30. The Schrage motor

The essential connexions of the Schrage motor, which is virtually an induction motor and a frequency changer combined into one machine, are given in Fig. 36. Contrary to the normal practice in induction motors, the primary winding is on the rotor and the secondary winding is on the stator. The commutator winding is sometimes called the *regulating winding*.

Arrangement of windings

In the Schrage motor, the primary winding performs two functions. It is the primary of an induction motor of which the stator winding is the secondary. It is also the primary of a frequency changer of which the commutator winding is the secondary. To a first approximation, therefore, the Schrage motor, like the 'induction motor-frequency changer' combination discussed in Section 29, is equivalent to an induction motor with a constant injected voltage.

The primary and commutator windings are wound in the same rotor slots, as in the typical section of a slot shown in Fig. 77. The diagram gives the usual arrangement with the primary winding at the bottom of the

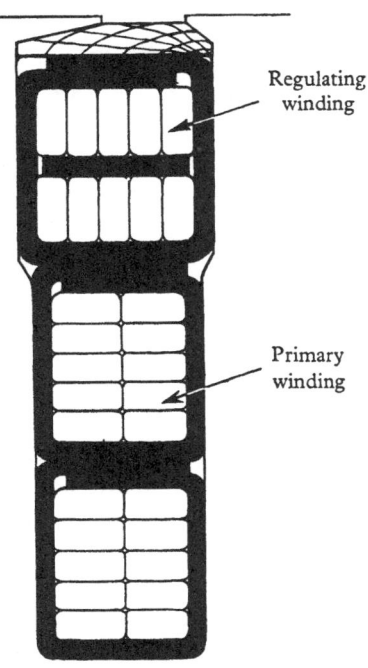

Fig. 77. Typical arrangement of rotor windings in the Schrage motor

slot. The reversed arrangement, with the regulating winding at the bottom, is sometimes used on very small motors in order to simplify the manufacture, but the arrangement shown in Fig. 77

is preferable from the point of view of the characteristics of the motor, and is assumed in the discussion which follows. It has the further advantage of improving the heat dissipation properties of the machine, because radial connectors, known as commutator risers, can be provided to assist the ventilation.

The rotor winding arrangement is the same as that of a frequency changer, in which separate slip-ring and commutator windings are provided, as described in Section 29. Theoretically a Schrage motor could be made with a single rotor winding, but this arrangement is never possible in practice because the commutator voltage has to be much lower than the voltage supplied to the primary winding.

Commutator voltage in the Schrage motor

The commutator of the Schrage motor provides the injected voltage required for controlling the speed and power factor. As explained on p. 38 in connexion with Fig. 26, the potential distribution set up at the commutator is constant, but when two sets of movable brush gear are fitted, the value of the output voltage of any phase depends on the positions of the two brushes belonging to that phase. The relationship can best be explained by means of the vector diagram of Fig. 78. The neutral brush position is defined in the same way as that of a frequency changer coupled to an induction motor.

If a single set of three-phase brush gear is used on a machine which has a constant flux, the equivalent star voltage per phase is constant in magnitude, but it varies in phase according to the angle of shift from the neutral position. Hence the locus of the end of the voltage vector is a circle as shown in Fig. 78, where OX is the voltage on neutral, and OA is the voltage when the brush gear is at an angle XOA away from neutral. The angles in the diagram represent electrical angles in the machine.

If two sets of brushes are provided, the voltage between a pair of brushes A and B belonging to one phase, which are occupying the positions corresponding to the points A and B in Fig. 78, is represented both in magnitude and phase by the vector AB. This diagram therefore represents not only the physical positions of the two sets of brush gear (when the angles are in electrical degrees) but also the polyphase voltage between them.

The magnitude of the voltage between brushes A and B depends on the angle $AOB = \theta$, irrespective of the actual position of the brushes. The angle θ is the angular distance in electrical degrees between the corresponding brushes of each phase, and is called the *brush separation*. The maximum voltage is called the *diametral voltage* and is obtained when θ is 180°. When θ is less than 180°, only that part of the commutator winding lying between the two brushes of the same phase carries the current of that phase.

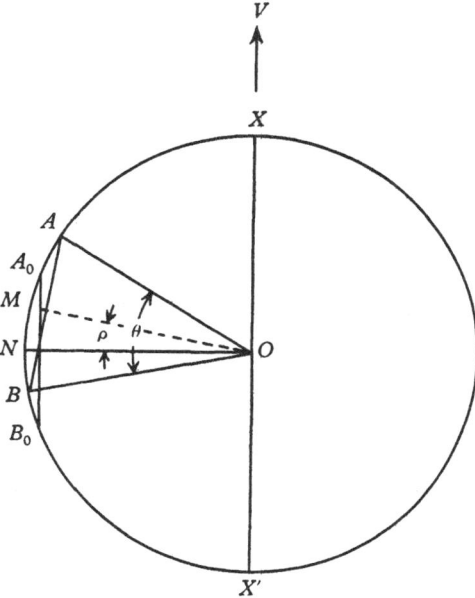

Fig. 78. Vector diagram of injected voltage
in the Schrage motor

Let the diametral voltage when referred to the primary circuit be E_d, and let the angle MON be ρ, where OM bisects the angle AOB and ON is perpendicular to XX'. The angle ρ is called the *brush shift*. The injected voltage referred to the primary is E_k, and the components of E_k are therefore

$$E_{kp} = E_k \cos \rho = E_d \sin (\theta/2) \cos \rho,$$
$$E_{kq} = E_k \sin \rho = E_d \sin (\theta/2) \sin \rho.$$

If the brushes are set at X and X' and are then moved in opposite directions by the same amount, the vector $A_0 B_0$ is always parallel to XX'; in other words, $\rho = 0$ whatever the brush separation may be. The motor is then said to be *operating on neutral*. Under this condition, the point N is always midway between A_0 and B_0. When the brushes are at A and B, where AB is not parallel to XX', the midpoint M is displaced from N by the angle ρ, which defines the phase of the voltage.

The angle θ is defined as positive when E_{kp} is positive (see p. 22), and hence a positive brush separation gives operation below synchronous speed and negative brush separation gives super-synchronous speeds. The angle ρ is positive when E_{kp} and E_{kq} have the same sign. In Fig. 78, the brushes have been shifted in the direction corresponding to a negative value of ρ. This is the direction which causes power factor improvement at sub-synchronous speeds, but at super-synchronous speeds it makes the power factor more lagging. A positive brush shift has the opposite effect.

In terms of the actual brush gear, a negative brush shift means a displacement of the brush gear in a direction opposite to the direction of rotation of the motor; this is known as *backward shift*. At speeds below synchronism, the flux in the Schrage motor rotates backwards in space, and hence a backward shift retards the phase of E_k. At super-synchronous speeds, the flux rotates forwards in space, and hence the slip-frequency voltage is advanced in phase if the brushes are shifted backwards. Since, however, the phase sequence in the secondary voltage diagram is reversed, as explained on p. 27, the angular movement of the vector representing E_k is still in a counter-clockwise direction. The angle of shift of E_k in the vector diagram resulting from a backward shift ρ is always therefore in the same direction, whether the speed is below or above synchronism.

A displacement of the brush gear in the same direction as the rotation is called *forward shift*. The effect of forward shift is exactly opposite to the effect of backward shift and corresponds to a positive value of ρ.

Brush mechanism

A typical example of the mechanism used for operating the brush gear of a Schrage motor is illustrated in Fig. 79. The parts of the commutator on which the two sets of brushes ride are separate, and are known as the *inner track* and the *outer track*. Each set of brushes is mounted on a steel brush ring which rotates in the end shield; in a machine without end shields, the brush gear is mounted in a separate yoke. On each brush ring there is an adjustable toothed rack engaged by a pinion mounted on a spindle. The two spindles pass through the end shield, and carry two

External pinions   Rack clamping bolt   Rack

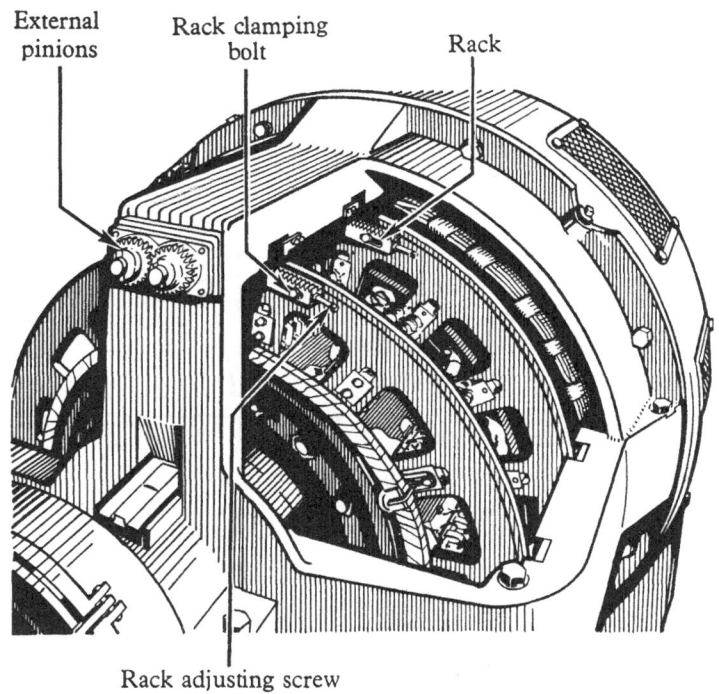

Rack adjusting screw
and lock-nut

Fig. 79. Arrangement of brush gear of Schrage motor

external pinions which engage together. In this way, the complete brush gear may be operated by rotating a hand-wheel or chain-sprocket mounted on the end of one spindle.

Normally the internal pinions and racks on the two rings are identical, so that equal movements of the spindles produce equal movements of the brushes. Hence, if the two external pinions are also equal, the brushes move equal amounts in opposite directions, and ρ is constant for all values of θ. If the initial setting is such that ρ is zero, the machine operates on neutral, that is, the injected voltage has a variable in-phase component E_{kp}, and zero quadrature component E_{kq}. Motors for reversing duty which have to operate alternately in both directions of rotation must operate on neutral, because any brush shift which improves the power factor in one direction of rotation makes it worse in the other. Such motors are therefore provided with equal external pinions.

Usually, however, the motor only runs in one direction, and the brush gear can be set so as to obtain the best results with that rotation. Shift of the brush gear can be used to obtain power-factor correction, and in general different values of shift are required at different parts of the range. The simplest and commonest method is to rotate the two brush rings at unequal rates by using external pinions with different numbers of teeth. A machine with this arrangement is said to operate with *varying shift*. The angle of shift ρ then changes as the separation is varied; a typical curve relating E_{kp} and E_{kq} is shown in Fig. 80. The brush shift can be adjusted in practice by loosening the racks from the brush rings and disengaging them from the internal pinions. Small adjustments of shift can be made, without disengaging the pinions, by using elongated bolt-holes in the racks. Once the brush shift has been set and the ratio of pinion teeth determined, it is unnecessary to alter the adjustment unless the motor is applied to a different duty, or has to be permanently reversed for any reason. For reversal of rotation the external pinions are interchanged and the brush shift adjusted.

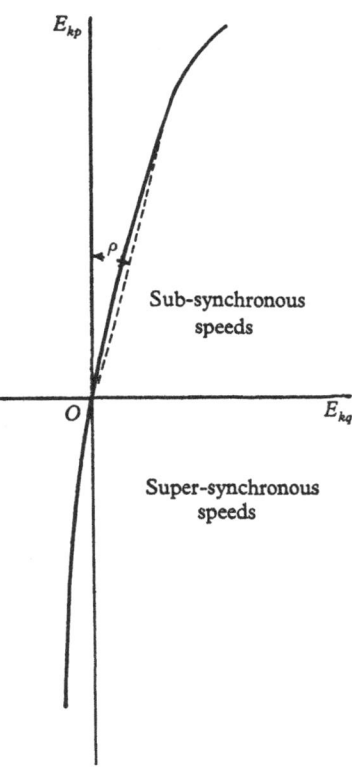

Fig. 80. Locus of E_k in the Schrage motor with varying brush shift

Range of speed variation

The speed of the machine is determined by the ratio of effective regulating winding turns to effective secondary winding turns. This ratio, called the *regulation ratio*, r, is approximately equal to the numerical value of E_k divided by the numerical value of E, and determines the slip of the motor at no-load. With the maximum brush separation, when the injected voltage is the diametral voltage E_d, r has the maximum value $r_{max.}$, and this determines the total range of speed obtainable from the motor.

The effective primary turns are fixed by the magnitude of the flux required, and the effective turns in the regulating winding by the permissible transformer voltage of commutation. But the number of turns in the secondary winding can be made whatever number is suitable, and therefore the desired value of $r_{max.}$ is obtained by choosing the appropriate number of secondary turns. If necessary, $r_{max.}$ can be greater than unity; under this condition, the motor can be slowed down to standstill on no-load with a brush separation less than the maximum of 180°, and further separation of the brushes causes the motor to run in the opposite direction of rotation, i.e. with slip greater than unity. It is uneconomical to design the motor with a greater regulation ratio than is needed, for decrease of secondary turns causes increase of secondary current, and therefore increase of brush gear and of commutator losses. Except in small sizes of machine, increase of speed range means increase in first cost and considerable decrease in efficiency at all speeds.

Starting

A Schrage motor is usually started up by switching directly to the supply when the brushes are in the low speed position. To ensure that the brushes are in this position a small interlock switch is fitted, connected in series with the low voltage release coil in the main supply switch; thus the main switch cannot be closed unless the brushes are in the low speed position. There are some exceptions to this practice. For instance, if the maximum regulation ratio is very small (say 0·2 or less) it may be necessary to insert secondary resistance at starting as in a slip-ring motor in order to reduce the starting current. The position of the brush gear is not then so important. In very small motors (say 2 H.P. and below) it is often permissible to start the motor with the brush gear in any position.

31. Theory of the Schrage motor

Before considering the vector diagrams of the Schrage motor, it is necessary to examine the effect of the inductive linking between primary and regulating windings. In the following discussion, the ratio of effective turns of the primary and secondary windings is assumed to be unity. All secondary quantities are therefore 'referred' values, as explained on p. 5. The regulating winding is understood to be that part of the whole commutator winding which is in action with the particular setting of the brush gear under consideration. The regulation ratio is r, and the angle of brush shift is ρ.

The total flux of the machine is considered to consist of four parts as indicated diagrammatically in Fig. 81. Although this shows only one slot on each side of the air-gap, it must be understood that the leakage fluxes link all conductors of the layer indicated. In addition to the four fluxes shown, there is a leakage flux linking the regulating winding only, but this is neglected because it has to cross two air paths and is therefore small.

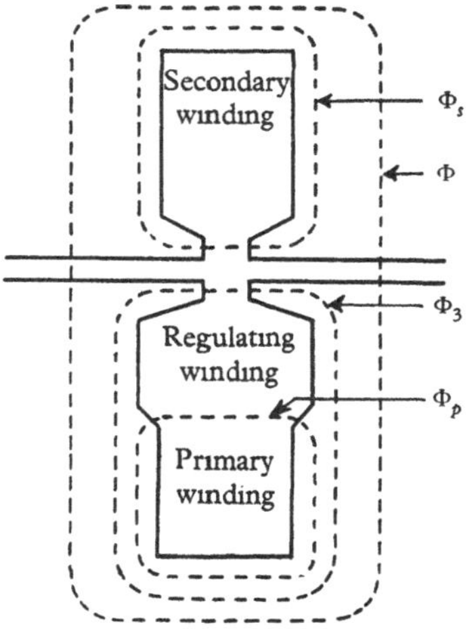

The four fluxes are:

(1) The main flux Φ, which links all three windings, and with which

Fig. 81. Separation of fluxes in Schrage motor

a corresponding magnetizing reactance X_0 is associated.

(2) The flux Φ_3, which links the two rotor windings only, and is a leakage flux due to the combined ampere-turns of the primary and regulating windings. It corresponds to a leakage reactance X_3. In the regulating winding there is, in addition to the voltages induced by the flux Φ_3, a component of leakage reactance drop caused by the change in current in the coils undergoing commutation. This is the second component $K_c n$ (see p. 48), and is neglected

in the theory because it is only a small proportion of the total regulating winding reactance, which is itself small.

(3) The primary leakage flux Φ_p, which links the primary winding only. It corresponds to the primary leakage reactance X_1.

(4) The secondary leakage flux Φ_s, which links the secondary winding only. It corresponds to the secondary leakage reactance X_s.

The total flux linking the two rotor windings is $\Phi_t = \Phi + \Phi_3$.

Let E be the voltage induced in the primary winding by the flux Φ, and E_t that induced by Φ_t. Then the primary voltage diagram is similar to Fig. 14, if E is replaced by E_t,

$$V = E_t + I_1 Z_1. \tag{54}$$

Fig. 15, the secondary voltage diagram, also holds, but in the secondary impedance $Z_2 = R_2 + jsX_2$, R_2 is the sum of the resistances of the secondary and regulating windings, and of the brushes. X_2, however, is the leakage reactance of the secondary winding only, because the whole of the induced voltage in the regulating winding is included in E_k. E_k is the voltage induced in the regulating winding by the flux Φ_t,

$$sE = I_2 Z_2 + E_k. \tag{55}$$

Also $$E = \frac{I_0}{Y_0}. \tag{56}$$

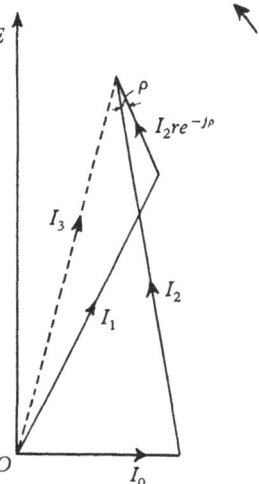

In the following analysis the core loss component G_0 of Y_0 is neglected, so that $1/Y_0 = jX_0$. The core loss can be included at the end of the calculation by adding a constant quantity to the friction loss.

Fig. 82 is a vector diagram of currents. It differs from Fig. 13 for the induction motor by the addition of the current $I_2 re^{-j\rho}$, which is a primary current producing the same M.M.F. as the current I_2 in the regulating winding.

Fig. 82. Vector diagram of currents in the Schrage motor

Then

$$I_1 = I_0 + I_2 - I_2 re^{-j\rho} = I_0 + I_2 K_2, \tag{57}$$

in which $$K_2 = (1 - re^{-j\rho}). \tag{58}$$

Equations (54) to (57) correspond to equations (8a) to (8c) and (20) for the induction motor. Further relations are required for E_k and E_t:

$$E_k = E_t re^{j\rho}. \tag{59}$$

E_t is the sum of E and a leakage component $jI_3 X_3$, where $I_3 = I_0 + I_2$.

Hence

$$E_t = \frac{I_0}{Y_0} + j(I_0 + I_2) X_3 = \frac{\sigma I_0}{Y_0} + jI_2 X_3, \tag{60}$$

in which

$$\sigma = 1 + \frac{X_3}{X_0} \tag{61}$$

is a real quantity slightly greater than unity.

The currents I_1 and I_2 and the torque can be deduced from these equations as follows:

Combining (55), (56), (59) and (60),

$$I_2 Z_2 = \frac{I_0}{Y_0} (s - \sigma re^{j\rho}) - jI_2 X_3 re^{j\rho}. \tag{62}$$

Combining (54), (57) and (60),

$$V = \frac{I_0}{Y_0} (Y_0 Z_1 + \sigma) + I_2 (K_2 Z_1 + jX_3). \tag{63}$$

Hence, putting

$$K_1 = s - \sigma re^{j\rho}, \tag{64}$$

$$c = Y_0 Z_1 + \sigma. \tag{65}$$

Equation (62) becomes

$$\frac{I_0 K_1}{Y_0} = I_2 (Z_2 + jX_3 re^{j\rho}). \tag{66}$$

Combining (63), (65) and (66),

$$K_1 V = I_2 [c(Z_2 + jX_3 re^{j\rho}) + K_1 (K_2 Z_1 + jX_3)].$$

Hence

$$I_2 = \frac{K_1 V}{K_1 K_2 Z_1 + cZ_2 + jX_3 (K_1 + cre^{j\rho})}. \tag{67}$$

Again, combining (54) and (60),

$$V = I_1 Z_1 + \frac{\sigma I_0}{Y_0} + jI_2 X_3. \tag{68}$$

Substituting I_0 from equation (57),

$$VY_0 = I_1(Z_1Y_0 + \sigma) - I_2(\sigma K_2 - jX_3Y_0)$$
$$= cI_1 - I_2[\sigma(1 - re^{-j\rho}) - (\sigma - 1)]. \qquad (69)$$

Hence
$$I_1 = \frac{1}{c}[VY_0 + I_2(1 - \sigma re^{-j\rho})]. \qquad (70)$$

The torque is calculated from the synchronous watts

$$W_T = mEI_2 \cos \phi_2, \qquad (71)$$

in which
$$E = \frac{I_1 - I_2K_2}{Y_0}, \qquad (72)$$

and ϕ_2 is the phase angle between E and I_2.

The complete characteristics of the Schrage motor can be calculated by means of equations (67), (70) and (71).

It may be noted that the vector ratios K_1 and K_2 are, in effect, transformation ratios for voltage and current respectively between the primary winding and the complete secondary circuit obtained by combining together the secondary and regulating windings.

32. Characteristics of the Schrage motor

Like the shunt motor, the Schrage motor possesses characteristics similar to those of an induction motor with constant injected voltage. A clear picture of the action of the Schrage motor can be obtained by thinking of an equivalent induction motor with a separate frequency changer. The total current, as shown in the vector diagram of Fig. 82, can be considered to be the sum of two primary currents, one corresponding to motor action and the other to frequency changer action.

The principal difference is that there is mutual coupling between primary and regulating windings. The effect is allowed for in the reactance X_3, and for accurate calculation it is important to include it.

Current locus curves

Since the equations of the Schrage motor (67) and (70) can be expressed in the form of equation (18) (p. 16), the locus of the primary current is a circle for any given brush separation and brush shift. The difference between the circles for the Schrage motor and those of Figs. 22 and 23 is caused by the fact that in the Schrage motor the frequency changer current increases the total primary current at super-synchronous speeds and decreases it at sub-synchronous speeds. This has a marked effect on the diameters of the circles; moreover, the point corresponding to

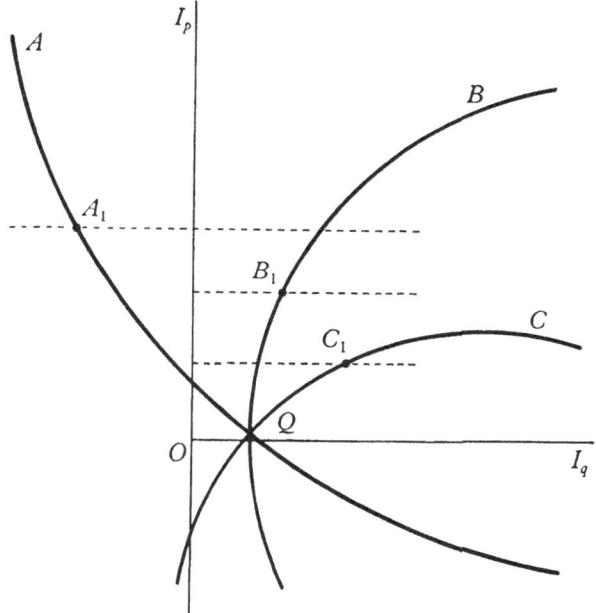

Fig. 83. Circle diagram of Schrage motor on neutral

full-load torque (assumed constant at all speeds) has a different component I_p for each circle instead of the constant component shown in Figs. 22 and 23.

Brushes on neutral

Fig. 83 shows three circles for the conditions when the motor operates below, near to, or above synchronous speed with the brushes set on neutral. The curves correspond to values $\theta = 180°$, $0°$ and $-180°$, with ρ zero throughout. This diagram should be compared with Fig. 22. Circle A for super-synchronous speed has a much larger diameter than circle B for synchronous speed, and circle C for sub-synchronous speed a much smaller diameter. It

was explained on p. 28 that the reversal above synchronous speed
of the secondary reactance vector in the voltage diagram causes
the secondary current to lead instead of lag, with the result that
at high speed (curve A) the Schrage motor has a high power
factor on full-load—often a leading one—even when the brush

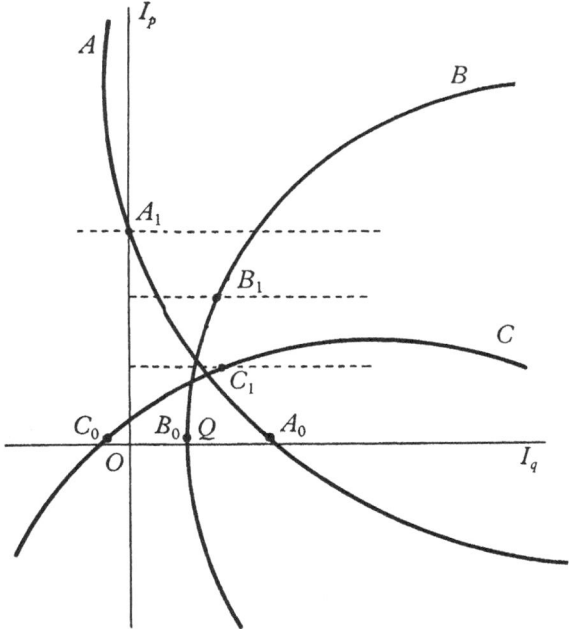

Fig. 84. Circle diagram of Schrage motor with varying brush shift

gear is on neutral and E_{kq} is zero. Since a leading power factor
involves a high secondary current, it is often desirable to give
brush shift at top speed in the backward direction, in order to
bring the power factor closer to unity.

Effect of brush shift

The circles of Fig. 84 show what can be achieved by suitable
varying brush shift. Circle B is unchanged, since this circle is
obtained when E_k is zero. Circle A is changed by a small backward shift while circle C is considerably altered. The backward
shift at this speed has not only moved the centre of the circle to

the left but has also increased its diameter. Thus all characteristics are improved. In particular, both primary and secondary currents are reduced at full-load, and as the bottom speed is usually the limiting condition from the point of view of heating owing to the reduced ventilation, this reduction of current is most important. As in the shunt motor, too much power-factor correction leads to excessive secondary current at light load; the brush shift at low speed is usually limited to that value which gives about full-load secondary current on no-load. With this amount of shift the primary current still lags at full-load, although it may lead slightly on no-load.

The circles of Fig. 84 should be compared with the corresponding circles of the shunt motor shown in Fig. 72. The comparison shows that at synchronous speed and below the power factor characteristic is better in the Schrage motor. This improvement is due to the fact that there is nothing corresponding to the constant regulator reactance in the secondary circuit. The small component $K_c n$ of commutator winding reactance, which, as explained on p. 152, is neglected in the theory, has a similar though much smaller effect. It causes the circles to slope slightly more to the right than those of Fig. 84, but experience shows that the effect is negligible in practice.

In the induction motor the diameter and position of the current locus circle is independent of the secondary resistance R_2. In the Schrage motor only the synchronous speed circle is independent of R_2. Since the brush-contact resistance in the Schrage motor forms a large proportion of the total secondary resistance, R_2 cannot be considered constant at all loads, for the contact resistance varies considerably with the value of current passing through the brush contact. Hence the true current locus is not a circle, and the circles found by using an average value of R_2 are only suitable for comparison with other machines. In practice an accurate determination of the characteristics can be made by estimating an appropriate value of R_2 for use in the calculation of each point.

Torque-speed curves

The torque-speed curves of the Schrage motor differ from those of the induction motor with constant injected voltage only in so far as the speed drop on load is increased. This increase is due to the addition of the resistance of the brush contact and the regulating winding. In general the drop in speed is greater than that of an induction motor and less than that of a shunt motor, particularly at low speeds. The variation of the phase of the secondary current at low speeds, combined with the secondary reactance, causes a relatively large drop in speed, which is, however, a good deal less than that of the shunt motor. In a Schrage motor with 3 to 1 speed range the full-load speed is about 20–30 % below the no-load speed.

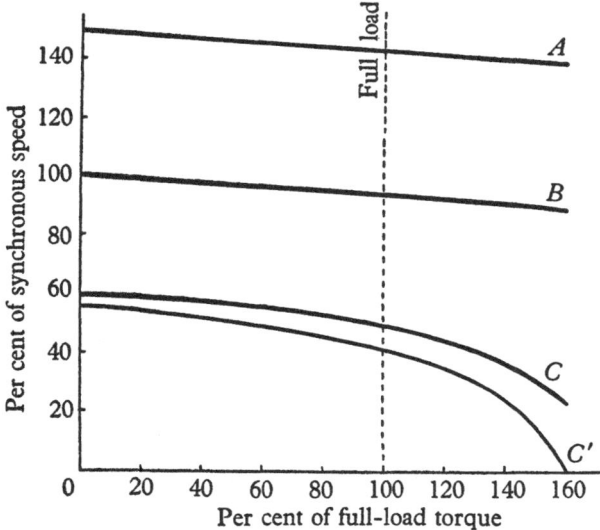

Fig. 85. Torque-speed curves of Schrage motor

Typical curves are shown in Fig. 85, which should be contrasted with Fig. 21. Curves A, B and C' apply to a machine with brushes on neutral. Brush shift has little effect on the curves at super-synchronous speeds, but a marked effect at sub-synchronous speeds in shortening the speed range and reducing the drop in speed from no-load to full-load. Curve C is the torque-speed curve with the same brush separation as that giving curve C' but with the brush shift corresponding to the current locus curves of Fig. 84. The starting torque is seen to be greatly increased by the shift given; although at the same time the speed range is somewhat reduced.

Like the current locus curves, the torque-speed curves are modified in practice because of the change in the brush-contact resistance as the load varies. This effect, which can be allowed for in the calculations by using appropriate values of R_2, does not greatly affect the general shape of the curves.

Control of the Schrage motor

The low-speed characteristic is important because it determines the starting conditions, when the motor is switched on to the supply with the brushes in the low speed position. The torque-speed curves C and C' of Fig. 85 show that the torque increases continuously as the speed is reduced, that is, there is no pull-out point in the region of positive speed. Hence a stable speed can be obtained at any load up to the stalling load, and during starting the acceleration is very smooth. The starting torque available depends on the speed range provided, and on whether brush shift can be given, that is, whether the motor has a reversing duty or not. It normally ranges from one and a quarter to twice full-load torque. The corresponding secondary current ranges from one and a half to three times the full-load value, but since owing to the frequency changer action, current is fed back to the supply, the primary current is less, and ranges from full-load to one and three-quarter times full-load current. Thus a high starting torque combined with a low starting current is obtained.

The smooth and accurate control obtained by moving the brush gear is of great value in many applications of the Schrage motor. The motor can act as a generator with any position of the brush gear, since the torque-speed curves, like those of an induction motor, can be extended to the negative side corresponding to generator action. Hence the machinery driven by the motor can be accelerated and retarded under the control of the brush operating mechanism. The pull-out torque as a generator is considerably higher than the maximum torque as a motor at low speeds, so that plenty of power is available for dynamic braking. On the other hand, the maximum generating torque at high speed is less than that as a motor.

Speed and brush separation

With the brushes on neutral and the motor on no-load, the relation between speed and brush separation is very nearly sinusoidal, because the relation between E_k and θ is sinusoidal. Since the speed variation from no-load to full-load increases at sub-synchronous speeds, the full-load speed does not vary sinusoidally with brush separation. Fig. 86 shows typical no-load and full-load curves indicated by full lines. If varying brush shift is used,

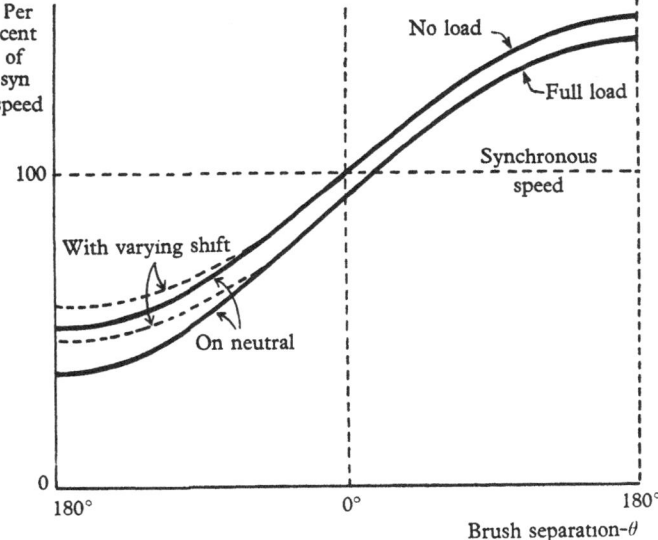

Fig. 86. Speed/brush separation curve for the Schrage motor

the speeds obtained with brush positions at the lower part of the range are increased. Typical curves for a motor with varying shift are shown dotted on Fig. 86.

The fact that more brush movement is required at the ends of the range for a given change of speed than in the middle is not usually a disadvantage, but there are some applications where it is necessary to have a linear relationship between brush movement and speed. It is then necessary to provide an increased amount of regulation, and use only the middle part of the curve where it is approximately straight.

Chapter 8

THE SCHERBIUS MACHINE

33. General

The Scherbius machine has already been defined as a polyphase A.C. generator, in which the output voltage is generated in the armature winding in a manner analogous to that in a D.C. generator. The action of the Scherbius machine is not limited to the positive generation of power, since it may also operate as a motor in the sense that power may flow into it from an outside source. However, the internal induced voltage is determined completely by the net exciting ampere-turns irrespective of the direction of power flow.

Self-excited low frequency generator

The use of a self-excited Scherbius machine as an independent low-frequency generator has been discussed in Section 21 and will not be considered further here. It should be noted, however, that although the shunt connexion of the exciting winding is similar to that of the shunt-excited speed-regulating machine discussed on p. 171, and the shunt phase advancer of Section 41, there is an important difference. The voltage and frequency of the auxiliary machines are greatly affected by the induction motor they control, while in the independent generator the voltage and frequency are determined only by the conditions in the machine itself.

Speed-regulating machine

The commonest use of the Scherbius machine is as a source of the injected voltage required to regulate the speed or power factor of a large slip-ring induction motor. The combination of an induction motor, a Scherbius machine and other associated apparatus is called a *Scherbius equipment*. Since the characteristics of the motor depend on the injected voltage, the control of the motor is a matter of controlling the excitation of the Scherbius machine. The number of possible combinations which have been used or proposed is very large and it is not possible to describe more than a few of the most important in this and the next chapter.

However complicated the complete scheme of the equipment may be, the excitation of the Scherbius machine must, as in a D.C. generator, be obtained in one or more of the following three ways:

(1) Series-excitation by a winding carrying the main current.

(2) Separate excitation derived from a source outside the Scherbius machine.

(3) Shunt-excitation from the terminals of the Scherbius machine.

Whatever method of excitation is used, the exciting current must be a polyphase current having the same frequency and phase sequence as the voltage which the machine is required to generate. With series and shunt-excitation this condition is automatically fulfilled. With separate excitation, special care must be taken to arrange the circuits so that the frequency is correct, and most of the methods of doing this involve the use of a frequency changer mechanically coupled to the main induction motor.

In the following three sections, various methods of Scherbius speed control of induction motors are considered according to which of the above three methods is used for the main excitation of the regulating machine. In each case, auxiliary excitation by one or both of the other methods may be used, too, in order to modify the characteristics. In Section 37 there is given a general theory which can be used to calculate the performance for any combination of the three methods of excitation.

Phase advancers

The Scherbius machine, as defined in Section 9, suitably proportioned so as to generate a relatively low voltage, can be used as a phase advancer for correcting the power factor of an induction motor without controlling its speed. Any of the three methods of excitation—series, shunt or separate—may be used depending on the characteristics required. These machines are described in Chapter 9.

It is desirable to clear up the confusion which exists about the nomenclature of these phase advancers. For instance, the simple machine with no stator winding dealt with on p. 47, and in Section 39, is called alternatively a *Scherbius phase advancer* or a *Leblanc exciter*. In this book the second of these names is used, because the machine does not come within the definition given here for a Scherbius machine. Also, the expression *Scherbius machine* is often considered to apply to the salient pole construction rather than to the arrangement of the windings as in the definition of Section 9. Phase advancers, especially the larger sizes, are sometimes made with the salient pole construction like speed-regulating machines, but more often the construction is simplified because of the low voltage (10–20 V.). Nevertheless these machines, both in theory and operation, come under the heading of Scherbius machines.

34. Series-excited speed-regulating machine

The simplest and earliest application of the Scherbius machine is the series-excited slip-regulating machine. This type is used with an induction motor for the purpose of increasing the slope of the torque-speed characteristic in applications where the motor has to carry heavy intermittent peak loads, as in certain rolling mill and winder drives. The motor is provided with a fly-wheel, which gives out some of its stored energy when the speed drops as the peak load comes on, thus reducing the peak torque demanded from the motor and the peak input from the supply system. A common requirement is that the slip of the induction motor shall be 10% at full-load and 20% at twice full-load. For such duties only a single characteristic is required, and no adjustable control is necessary.

A characteristic of this type can be obtained by connecting a fixed three-phase resistance in the rotor circuit of the motor. The torque-speed characteristic of this combination is described on p. 19, and shown in Fig. 12. A series-excited Scherbius machine gives a similar result with the advantage that some of the power which would be lost in the resistance is saved. There is also the possibility of correcting the power factor.

Series slip regulator without power-factor correction

A typical diagram of connexions is given in Fig. 87. Here the Scherbius machine, together with an A.C. motor to which it is coupled, form a separate set which is mounted independently of the main motor and runs at an approximately constant speed determined by the driving motor. When the main motor is loaded, the slip regulator takes electrical power from the rotor, and the

A.C. driving motor, acting as a generator, returns most of this power to the supply. The exciting winding carries the secondary current of the main induction motor, which is also the main current of the Scherbius machine.

When the exciting winding is arranged as in Fig. 42, and the regulating machine is unsaturated, the induced voltage is in phase with and proportional to the secondary current. The regulating

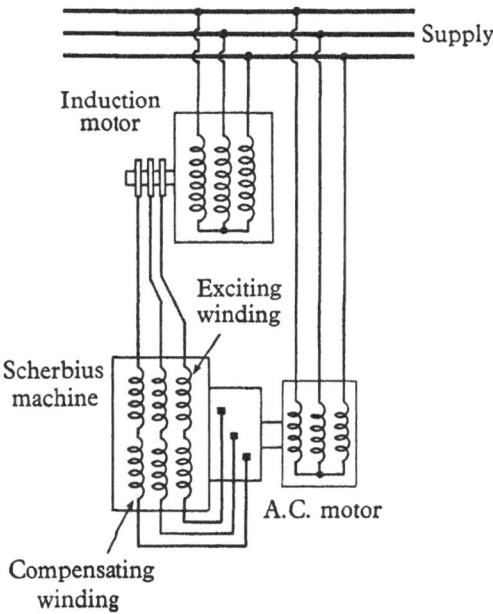

Fig. 87. Connexions of series
slip regulator

machine is then equivalent in this respect to a resistance, and the characteristics of the main motor are the same as those of an induction motor with a fixed secondary resistance. The current locus is the ordinary circle diagram shown in curve 1, Fig. 88, and the torque-speed curve is curve 1 a, Fig. 89 (full line).

If the regulating machine becomes saturated at higher values of excitation, the voltage it provides is less than that given by the secondary resistance. A similar result is obtained if the regulating machine is coupled to the main motor instead of to a separate A.C. motor. Although this arrangement is more efficient and sometimes more economical, it has the effect of reducing the injected voltage at high loads because of the fall in speed. The result is to change the torque-speed curve to that shown dotted (curve 1 *b*) in Fig. 89, which is inferior to curve 1 *a* from the point of view of utilization of fly-wheel energy.

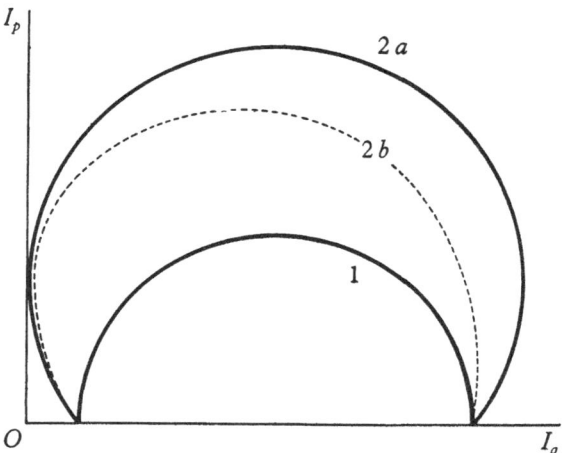

Fig. 88. Current locus diagrams of induction motor with series slip regulator

Series slip regulator with power-factor correction

By suitable interconnexion of the phases of the exciting winding, it is possible to make the induced voltage lag behind the main current of the Scherbius machine and hence to correct the power factor of the induction motor as well as to increase its slip. For example, if the current of phase A passes through a coil on pole B in Fig. 42 in series with coil A (and similarly for the other phases) the phase angle between voltage and current will depend on the ratio of turns in the two coils.

If the injected voltage is proportional to the current and at a constant phase angle with it, the current locus curve of the induction motor becomes a circle whose centre is raised above the axis, as shown in Fig. 88, curve 2 *a*; the torque curve tends to curve upwards compared with that obtained by resistance (Fig. 89, curve 2 *a*). This result can be derived either geometrically or by the analytical method to be given in Section 37.

As before, the effect either of saturation in the Scherbius machine, or of the loss of speed at higher loads when it is direct coupled to the induction motor, is to distort the characteristics as shown by the dotted lines, curves 2*b* in Figs. 88 and 89. The resulting torque curve may therefore be considerably inferior to that obtained with a slip resistance.

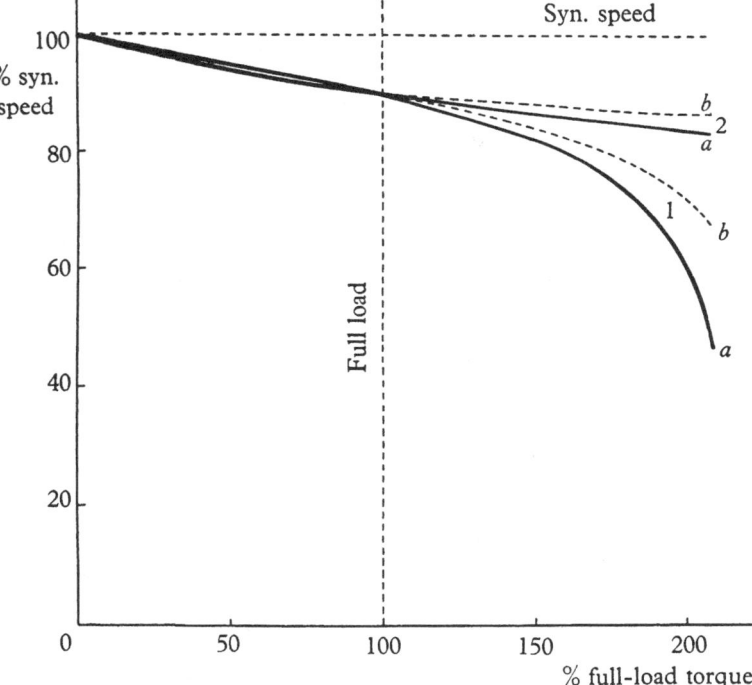

Fig. 89. Torque-speed curves of induction motor with series slip regulator

Many equipments of this type were built in the early days of electric motor drives, but the application to this duty has been much less in recent years. The use of a slip-regulating machine of this type is only justified if the load on the motor is well maintained, since the saving of power and the correction of the power factor both depend on the load. At light loads, the power factor is little improved, and the total losses may actually be higher than when running with resistance. If the motor runs with a low load-factor, the total saving does not justify the cost of the additional gear.

Series slip regulator with additional separate excitation

Better results can be obtained if, in addition to the main series winding, the Scherbius machine has an auxiliary exciting winding separately excited from a frequency changer exciter driven directly by the main motor. The separate excitation is arranged to give power-factor correction at no-load, and also to give a no-load motor speed above synchronism, thus reducing the size of the slip-regulating machine, and compensating for the extra cost of the exciter.

35. Separately excited speed-regulating machine

An induction motor with a series-excited speed-regulating machine connected in the secondary circuit, has a single torque-speed characteristic which is not varied during operation. A more frequent requirement is a motor with an adjustable speed which is variable over a certain range by means of a controlling device. It is also generally desirable that the motor shall have a shunt characteristic. These requirements can be met for large and medium outputs by using an induction motor with a speed-regulating Scherbius machine having either separate or shunt-excitation or a combination of the two.

The most straightforward method is to use separate excitation obtained from a frequency changer; the speed of the main motor can then be controlled either by varying the voltage supplied to the frequency changer, or, as in the Schrage motor, by moving two sets of brushes on the commutator of the frequency changer. There are, however, certain difficulties, mainly due to the fact that, while the reactance of the exciting circuit varies with the frequency, at the same time it is necessary to control the phase of the injected voltage with considerable accuracy. As a result, successful application of this method has been limited to motors with relatively narrow speed ranges. Where, however, the lowest speed is not less than 80% of the maximum speed, this type of equipment provides efficient and fine speed control, and is particularly useful if automatic control is necessary.

Scherbius equipment with brush-shifting exciter

Fig. 90 shows the main connexions of an equipment in which the frequency changer has two sets of movable brushes. The

Scherbius machine is here coupled to the motor, but may equally well be separately driven. If the induction motor is supplied at high voltage, a transformer is required to supply the slip-rings of the frequency changer. The Scherbius machine has two exciting windings. The main exciting winding is connected in series with a resistance between the two brush sets on the frequency changer, and it provides the main excitation required for speed regulation. The auxiliary winding is connected in star and is supplied from one brush set only, so as to maintain a small excitation for power-

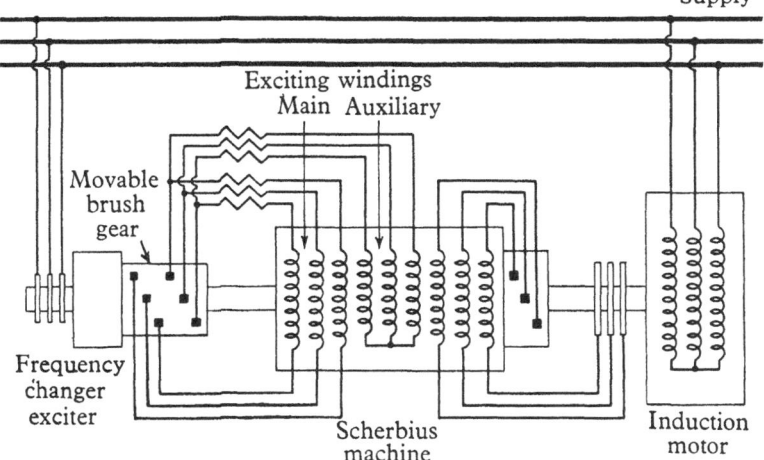

Fig. 90. Scherbius equipment with brush-shifting exciter

factor correction when the two brush sets are coincident and provide no main excitation. The auxiliary winding also has a resistance in series.

The control of speed of this equipment is similar to that of the Schrage motor, but additional precautions are necessary to control the phase of the injected voltage because of the effect of the reactance of the exciting winding. At very low secondary frequencies, the exciting current in each winding is almost in phase with the voltage applied to it, but as the slip of the motor increases above or below synchronism, the current lags by an increasing amount. The resistances are used to limit the phase displacement to a reasonable amount. It is the loss in the resistances which sets a limit to the speed range obtainable with this type of equipment.

In order to control the phase of the injected voltage and thus keep the power factor of the motor within reasonable limits, it is generally necessary to use a cam mechanism for operating the exciter brush gear. If this is suitably designed, the power factor of the motor can be maintained close to unity at all speeds. Typical torque-speed curves are shown in Fig. 91, in which the drop in speed from no-load to full-load is only a few per cent.

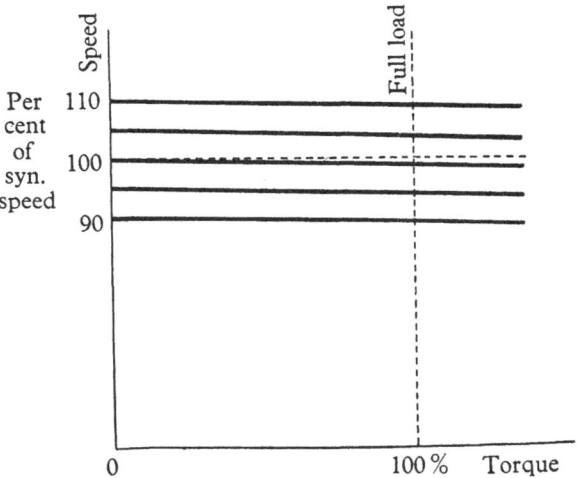

Fig. 91. Torque-speed curves of Scherbius equipment with brush-shifting exciter

Other methods of control

Another method of control uses a frequency changer with fixed brushes, but with its slip-rings supplied from two double induction regulators, one to control the speed of the main motor and one to control its power factor. In yet another method the frequency changer is supplied from a special alternator driven by a synchronous motor. The alternator has distributed exciting windings arranged so that its voltage can be controlled both in magnitude and phase. In some large installations additional machines or transformers have been introduced in the exciting circuits in order to modify the characteristics according to particular requirements.

Additional series-excitation

In addition to the separate excitation, a small amount of series-excitation may be used on the Scherbius machine in order to modify the characteristics. Some series-excitation is always present, because of errors of compensation, and because the brushes must always be slightly forward from neutral in order to avoid the danger of instability, due to self-excitation (see p. 111).

36. Shunt-excited speed-regulating machine

The best known and most widely used type of Scherbius equipment has a regulating machine with shunt-excitation, either with or without a small amount of separate excitation obtained from a frequency changer. If shunt-excitation only is used, the speed regulation is limited to the sub-synchronous range and the machines form a *single-range* equipment. By using in addition a frequency changer coupled to the main motor, the motor speed can be regulated above synchronism as well as below, and the set then becomes a *double-range* equipment. In both cases the motor has shunt characteristics.

Single-range Scherbius equipment

Fig. 92 shows a typical connexion diagram of a single-range equipment, in which the Scherbius machine is driven by a separate A.C. motor. The exciting winding is supplied in shunt from the terminals of the Scherbius machine, through a transformer (usually an auto-transformer) with a number of taps. Speed control is obtained by changing the taps by means of special control gear. The phase connexions of the exciting winding and transformer are made in such a way that the voltage induced in the Scherbius machine is approximately in phase with the secondary induced voltage of the induction motor, that is, the phase of the injected voltage is such as to produce speed variation.

The general action of the machine is as follows. The current passed through the exciting winding by the voltage applied to it, must induce an armature voltage such that the terminal voltage is equal to that applied to the exciting winding, as in any self-excited machine. The conditions differ, however, from those in a D.C. shunt generator because of the reactive drop in the exciting winding, which is normally much greater than the ohmic drop.

The balance is therefore mainly between the armature induced voltage and the reactive induced voltage in the exciting winding. Since the reactive voltage leads the exciting current by 90°, a phase displacement of approximately 90° is necessary either between the primary and secondary voltages of the transformer, or between the exciting current and the armature induced voltage.

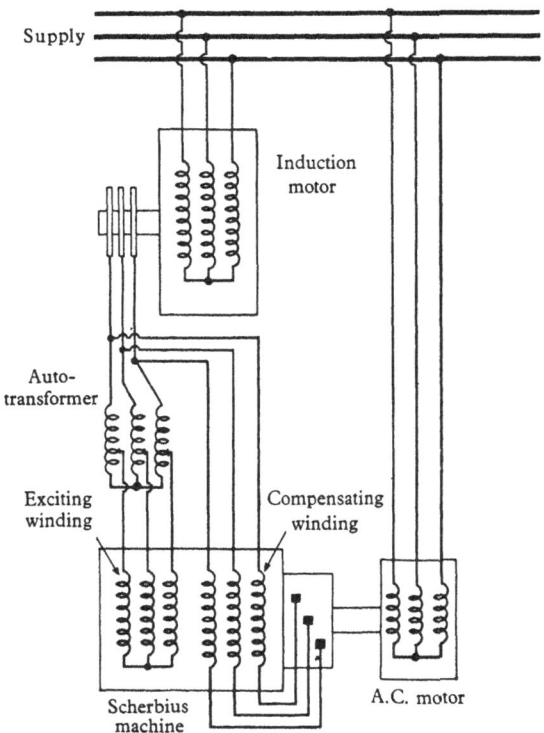

Fig. 92. Single-range Scherbius equipment

If a two-winding transformer is used, a 90° phase displacement between primary and secondary voltages can be obtained by means of a delta-star connexion. The exciting winding is then arranged as in Fig. 42, so that the armature voltage is in phase with the exciting current. Usually, however, an auto-transformer is used, and the 90° displacement is obtained by making each phase of the exciting winding consist of two coils in series, one on each of the poles corresponding to the other two phases. Thus exciting phase A consists of two equal coils on poles B and C, and hence the armature induced voltage in phase A is 90° out of phase with the exciting current in phase A. An auto-transformer is more economical and efficient than a two-winding transformer, because a unity ratio of transformation can be used when the maximum slip is required on the induction motor. For speeds nearer to synchronism a fractional tap is used.

The voltage induced in the armature and compensating windings together, as shown on p. 49, equation (31 c), is

$$E = \pi\sqrt{2}\Phi T_e n(p/2)\, 10^{-8}, \qquad (73)$$

while that in the exciting winding is

$$e = \pi\sqrt{2}\Phi f T_x 10^{-8}, \qquad (74)$$

where Φ = flux per pole,

   f = frequency,

   n = speed in rev./sec.,

   T_e = effective number of armature turns,

   T_x = effective number of exciting turns.

If the impedance drop in the main windings and the ohmic drop in the exciting winding are neglected, the ratio of e to E is the transformation ratio K_T of the transformer. Hence

$$f = K_T n \frac{p}{2} \frac{T_e}{T_x}. \qquad (75)$$

The frequency is therefore determined approximately by the connexions of the exciting circuit, and apart from saturation the Scherbius machine could operate at any value of voltage. In practice, the voltage is determined by the induction motor, because there is a definite relation between the secondary voltage and frequency. Thus the speed of operation of the motor is that corresponding to the secondary frequency given by equation (75), and the Scherbius machine has to provide whatever value of injected voltage is required for this speed. Speed variation is obtained by adjusting the transformer ratio K_T by means of the tap-changing gear.

The above result is only approximate because both the impedance drop due to the load current and the resistance drop in the exciting winding have been neglected, as well as the impedance and magnetizing current of the transformer. Moreover, it is generally desirable to correct the power factor of the induction motor above the value obtained when it operates alone. For all these reasons the phase displacement between the exciting current and the armature induced voltage must in practice be less than 90°. To bring this about, additional windings are provided on the transformer and connected in series with the exciting winding so as to introduce a small component of voltage (usually about 20%) in quadrature with the main exciting voltage obtained from the secondary side of the transformer. These auxiliary sections are in circuit with all positions of the tapping switch, and for fine adjustment of the power factor, a small resistance is connected in series with each tap. The auxiliary transformer sections and the resistances are not shown on Fig. 92.

Characteristics of single-range equipment

Because the Scherbius machine tends to maintain a constant secondary frequency, and hence a constant motor speed, the motor has shunt characteristics in which increase of load causes only that small decrease in speed due to secondary resistance drop. Because of its dependence on shunt-excitation, however, the single-range equipment gives some interesting results, particularly when operating near to synchronous speed.

In the first place, it is evident that if the motor is running unloaded at synchronous speed, with zero secondary voltage and current, it will continue to do so whatever changes are made in the control apparatus. Hence all the torque-speed characteristics pass through the point corresponding to synchronous speed and no-load, and the speed cannot be regulated if the motor is unloaded.

Fig. 93 shows some torque-speed curves of the kind obtained from a single-range Scherbius equipment. From these curves, it can be seen that once the speed has been brought down it will stay there. If properly designed, a single-range equipment can be brought down from synchronous speed if the load is not less than about 20% of full-load, by passing successively from one tap to the next. It is not suitable for drives where the motor may be completely unloaded or where generator operation is required.

The current locus curves (Fig. 94) also must pass through the point corresponding to zero secondary voltage, that is, the point A. The four curves in Fig. 94 are marked with the same numbers as the corresponding torque-speed curves of Fig. 93. The curves are not circles but are higher order curves which loop round and ultimately come back to the point A.

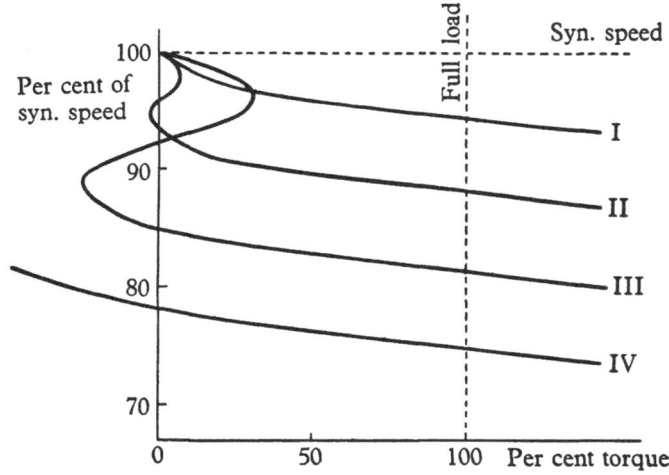

Fig. 93. Torque-speed curves of single-range Scherbius equipment

Double-range Scherbius equipment

The equipment illustrated in Fig. 92 is only suitable for single-range operation below synchronous speed. However, in order to make full use of the equipment, it is desirable to be able to regulate the speed of the motor above synchronism as well as below. Although the equipment in Fig. 92 would operate above synchronous speed if the transformer voltage were reversed, it is not possible to raise the speed *through* synchronous speed because at that point the frequency is zero and the transformer therefore cannot provide any excitation for the regulating machine. No practical scheme has yet been evolved for raising the speed of an induction motor through synchronous speed except with the help of a frequency changer. Owing to the fact that it can still generate a voltage even at zero frequency, the frequency changer, either as a direct regulating machine or as an exciter for a Scherbius machine, can be used to regulate the speed of an induction motor above synchronism.

In a double-range shunt-excited Scherbius equipment, of which the main connexions are given in Fig. 95, the regulating machine

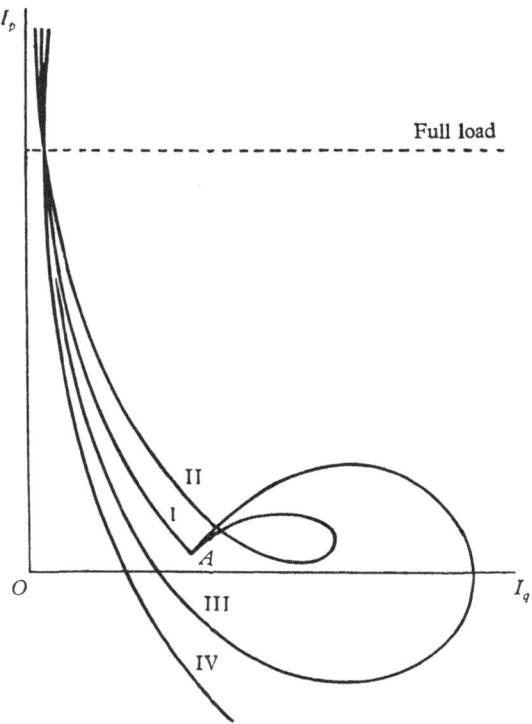

Fig. 94. Current locus curves of single-range
Scherbius equipment

has, in addition to the shunt-excitation from the auto-transformer, a small amount of separate excitation from a frequency changer exciter, direct coupled to or driven by gears from the main motor shaft. In this way a wider range of speeds above as well as below synchronous speed can be obtained if the same size of regulating machine is used, or alternatively, the same speed range can be obtained using a smaller machine.

For large equipments the double-range scheme is more economical than the single-range scheme, in spite of the additional complication, and it has been used to a consider-able extent for variable-speed rolling mill and other drives. It involves, however, the use of extra reversing contactors and other auxiliary apparatus in addition to the frequency changer,

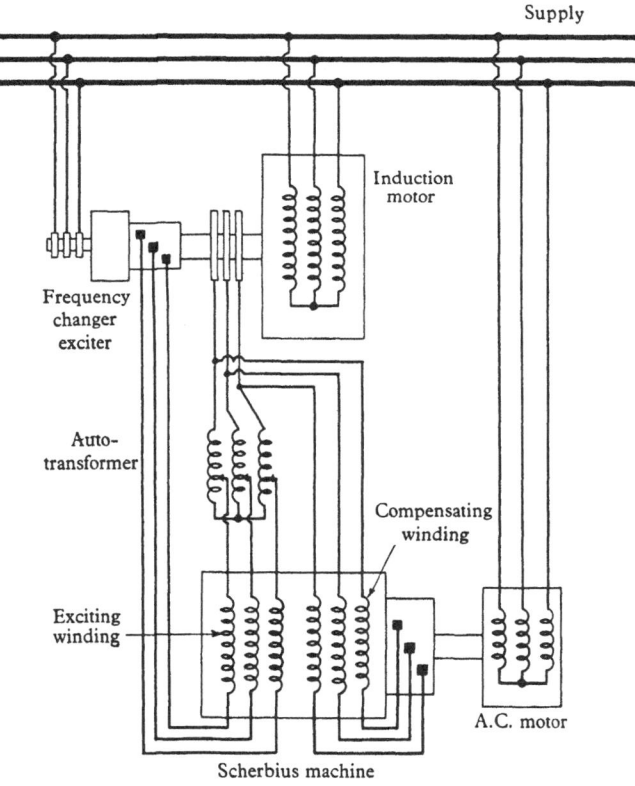

Fig. 95. Double-range Scherbius equipment

and for smaller horse-powers, particularly when only a short speed range is necessary, the simpler single-range equipment may have the advantage.

As in a Scherbius machine with separate excitation, there may be some series-excitation, due either to errors of compensation, or to a series winding provided for the purpose of modifying the characteristics.

37. Calculation of Scherbius equipments

It has been explained how a wide range of requirements can be met, when large powers are involved, by means of various types of Scherbius equipment, in which the excitation of the Scherbius machine is a combination of series, separate and shunt-excitation.

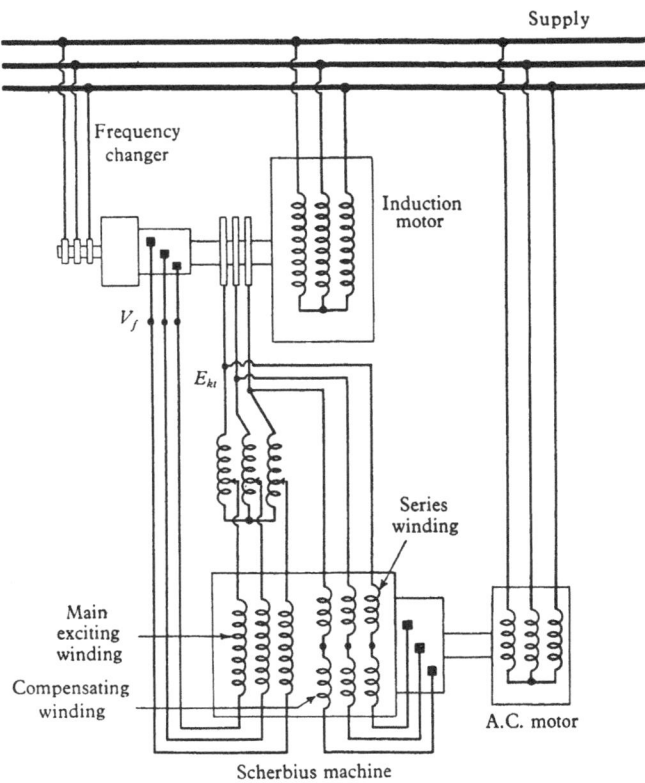

Fig. 96. Scherbius equipment with three
methods of excitation

In order to obtain a general theory, an equipment, in which all three methods of excitation are used at the same time, is considered. The diagram of connexions is that shown in Fig. 96. The complete diagram would apply, for example, to a double-range Scherbius equipment having, in addition to combined shunt and separate excitation applied to the main exciting winding, a small amount of series-excitation arising from inexact compensation. For equipments in which only one or two of the three methods of excitation are used, it is only necessary to omit the appropriate terms from the equations.

The main circuits of the induction motor and the Scherbius machine are represented by the equivalent circuit of Fig. 97, which is like that of Fig. 16, except that some other components are added. It is not feasible to include the main exciting winding in a single comprehensive equivalent circuit, and it is preferable to deal with the exciting circuit by means of an equation relating

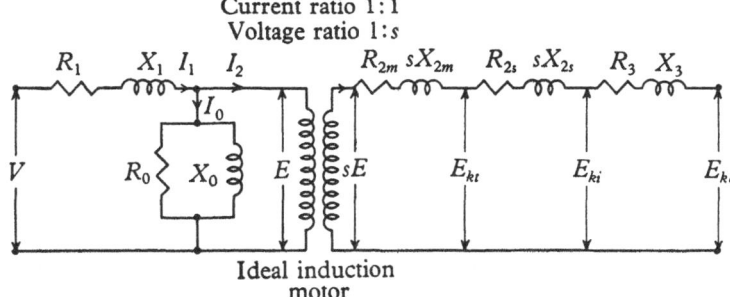

Fig. 97. Equivalent circuit of Scherbius equipment with three methods of excitation

the quantities shown in Fig. 97. By combining the exciting circuit equation with those associated with the main equivalent circuit a complete solution is obtained. It is assumed that the Scherbius machine is unsaturated, and that the speed is given. In Fig. 97 the circuit to the left of the ideal motor is the same as in Fig. 16:

$Z_{2m} = R_{2m} + jsX_{2m}$ is the secondary impedance of the motor.

$Z_{2s} = R_{2s} + jsX_{2s}$ is the impedance of the main circuit of the Scherbius machine. There is no constant component of reactance ($K_c n$, see p. 48) if an interpole is used to neutralize the voltage induced in the short-circuited coil.

E_{kt} is the secondary terminal voltage, and corresponds to the injected voltage E_k of Fig. 16.

E_{ki} is the internal induced voltage of the Scherbius machine, resulting from the combined action of the main exciting winding and the series winding.

$Z_3 = R_3 + jX_3$ is a fictitious impedance representing the effect of the series winding, since the voltage due to a series current is proportional to and at a constant phase angle to the current.

E_{ke} is the voltage which would be induced in the armature by the main exciting current I_e if it acted alone.

The exciting current in the main exciting winding is small in comparison with I_2, and it is therefore assumed that the current in Z_2 and Z_3 is also I_2.

If there is no saturation, the total induced voltage E_{ki} can be considered, by the principle of superposition, to be the resultant of the voltage E_{ke} due to the main exciting current, and the voltage $I_2 Z_3$ due to the series current. Hence

$$E_{ki} = E_{ke} + I_2 Z_3, \tag{76}$$

and
$$E_{kt} = E_{ki} + I_2 Z_{2s}. \tag{77}$$

The current I_e in the main exciting winding is proportional to E_{ke}. There are four voltages in the exciting circuit:

V_f is the constant voltage from the frequency changer which provides the separate excitation.

$K_t E_{kt}$ is the shunt-excitation voltage obtained from the transformer.

$I_e R_e = K_e E_{ke}$, the resistance drop in the exciting circuit (including the resistances of the transformer, frequency changer and any resistors). The leakage reactance of the exciting winding is neglected.

$sK_2 E_{ki}$ is the voltage induced in the main exciting winding by the resultant flux of the Scherbius machine.

The exciting circuit equation is therefore

$$V_f + K_t E_{kt} + sK_2 E_{ki} + K_e E_{ke} = 0. \tag{78}$$

The quantities V_f, K_t, K_2 and K_e are constant vectors because the relationships they express include a constant phase angle as well as a constant magnitude.

V_f depends on the brush position on the frequency changer as well as on its transformation ratio.

K_t depends on the vector resultant of the voltages of the main and auxiliary sections of the transformer, and on the location of the exciting winding on the poles of the Scherbius machine.

K_2 and K_e depend on the location of the exciting winding.

All the above values are referred to the primary of the induction motor in the same way as the voltage, current and impedance of the induction motor secondary. This is generally the most convenient method, because, although it involves some initial computation, it makes the working, and any vector diagrams based on the calculations, much clearer. As before, the quantities relate to one primary phase.

Equations (76), (77) and (78), in combination with those given on pp. 22–4 for the induction motor, can be used to calculate the primary current and the torque of the motor for any value of slip. It is not usually worth while to derive a single expression for I_1, but it is preferable to obtain it by means of a step-by-step vector calculation. The work involved is then not excessive even for a complicated equipment.

Alternatively the equations can be used for calculating the constants of the transformer and the frequency changer required to give specified input or load conditions at a given value of slip. This leads to a straightforward method of determining the design of the control apparatus required for a specified range of duty.

The equations given above do not cover completely all possible types of Scherbius equipments, because other elements, such as additional windings, machines or transformers are frequently added in order to modify the characteristics. The general effect is to modify the exciting circuit equation without affecting the main circuit. There is no difficulty in introducing additional terms in the exciting circuit equation to allow for the added apparatus, and, if this is done, the method explained can be used to deal with any combination which is likely to arise.

Characteristics

The wide range of possible methods of excitation in Scherbius equipments allows great flexibility in the control of the injected voltage applied to the induction motor. With suitable control apparatus almost any desired characteristics can be obtained. The principal limitation is that the speed range obtainable is relatively short, as it is not usually possible to operate with a secondary frequency above about 15 cycles per second.

In general, a Scherbius equipment with predominantly shunt or separate excitation operates with shunt characteristics and can be designed to have a good power factor under all conditions of operation. Variation of the speed is obtained by controlling the excitation. On the other hand, a series-excited Scherbius machine causes the slip of the induction motor to vary with the load, and can be used as a rotary slip regulator for fly-wheel drives.

Chapter 9

PHASE ADVANCERS AND
COMPENSATED MOTORS

38. Commutator machines for power-factor correction

It has already been mentioned on p. 26 that a commutator machine may be used for correcting the power factor of an induction motor without greatly affecting its speed, and is then called a *phase advancer*. In a similar way a self-contained commutator motor may be arranged so that power-factor correction is obtained, but no speed variation. Such a motor is known as a *compensated motor*.

The injected voltage required to correct the power factor of an induction motor without affecting its speed is relatively small, amounting generally to a few per cent of the standstill secondary voltage. Hence the phase advancer, which carries the full secondary current of the induction motor, generates only a low voltage, usually between 10 and 20 V. It is therefore a small machine with a large commutator, carrying a heavy current in relation to its size. In a compensated motor the windings which generate the injected voltage are of small capacity, and the commutator, although large in relation to the power it handles, is small compared with that of a variable speed motor.

Characteristics

The characteristics of an induction motor with a phase advancer depend on the type of machine used. The torque-speed curve is essentially the same as that of the induction motor by itself and need not be considered further. The important characteristic is the current locus curve, which shows what power factor is obtained and how this varies with the input. Some types of phase advancer are set without means of adjustment, while, with others, the power factor can be varied by shifting the brushes or by operating a rheostat. In general the effect of the phase advancer is to move the working part of the circle diagram of the induction motor (see Fig. 10) to the left.

The various types of phase advancer and compensated motor can be considered under the three headings used in Chapter 3 for the general classification of commutator machines. There are, however, in addition, two types of phase advancer which do not fall into these three groups, namely the Leblanc exciter and the Kapp vibrator, and five headings are necessary to cover the entire field, as summarized in the next paragraphs. In Sections 39–42 of the present chapter, the more important types are described in more detail.

Leblanc exciter

The Leblanc exciter consists of a polyphase commutator armature rotating in its own field, as explained on p. 47. A compensated motor formed by combining together an induction motor and a Leblanc exciter into a single machine is called the Torda motor and it has been made under that name on the Continent, and in England under the trade name of 'All-watt' motor.

Shunt motor

The shunt motor offers the possibility of obtaining a compensated motor operating with an approximately constant speed at varying loads. Such motors have been built, but they are uneconomical because the size of the commutator can only be reduced by increasing the induced voltage in the commutator winding at standstill. Thus the commutator is large unless special starting gear is provided.

Commutator frequency changer

The frequency changer phase advancer gives very good operating characteristics, but it is limited to special applications, owing to the fact that it must be mechanically coupled or geared to the main motor and also because the number of poles is determined by the speed. On the other hand, the Osnos compensated motor, which is a modification of the Schrage motor, has been built extensively on the Continent, and also under the trade names of 'No-lag' and 'Kosfi-leading' motors, in England.

Scherbius machine

The most popular types of phase advancer are modifications of the Scherbius machine, either with series or shunt-excitation.

These machines have the advantage that they are independent of the main induction motor and can be driven at any convenient speed, so that some standardization is possible.

Kapp vibrator

The Kapp vibrator consists of three separate D.C. machines, generally connected in delta, and mounted in a common frame. The three points of the delta are connected to the slip-rings of the induction motor. The armatures do not rotate continuously, but they run up, stop, and reverse in time with the alternation of the current, and by a combined electrical and mechanical action, generate a voltage which lags behind the current. The Kapp vibrator, which is seldom built now, is not strictly a polyphase A.C. machine.

39. The Leblanc exciter

The simplest type of phase advancer is known as the Leblanc exciter, and it consists of a polyphase commutator armature rotating in its own field. It is driven at a suitable speed by a small motor or by belt-drive from the main motor. The polyphase armature currents set up a rotating flux which induces polyphase voltages of the same frequency between the brushes. As shown on p. 47 the induced voltage is in quadrature with the current and lags behind the current if the armature runs in the same direction as, and faster than, the field.

Two different constructions are used for this type of machine. In the first arrangement, the machine, often called the 'Scherbius

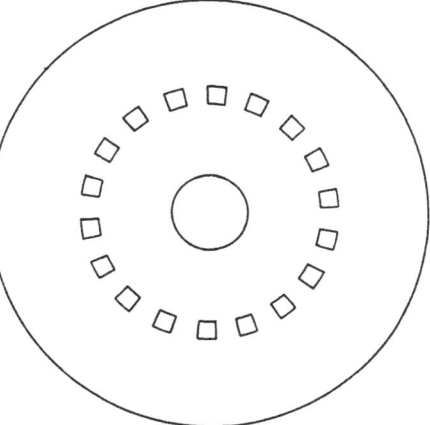

Fig. 98. Punching of Leblanc exciter showing winding slots with single core

phase advancer' (see p. 163), consists of a single laminated core in the middle of which are equally spaced holes completely surrounded by iron (see Fig. 98). The bars which form the armature winding are pushed through these holes, and are connected to the commutator. No stator is required, as the magnetic circuit linking the winding is completed inside the rotating core.

In the second arrangement, separate stator and rotor cores with an air-gap between them are used. Since there is no stator winding, the two parts may rotate together, provided the air-gap is maintained, but the usual practice is to mount the outer laminations in

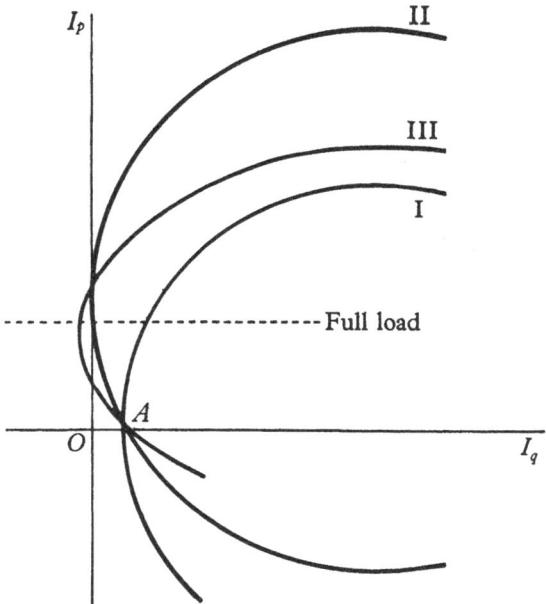

Fig. 99. Current locus curves of induction motor
with Leblanc exciter

a stationary frame as in an ordinary machine. One advantage of using an air-gap is that the self-inductance of each individual coil is reduced, and commutation is improved. Another is that interpole windings can be provided on the stator for assisting commutation. Apart from the question of commutation the characteristics of the two types are the same.

As shown on p. 166, the effect of injecting a voltage which is proportional to the current and lags behind it at a constant angle, into the secondary circuit of an induction motor, is to bulge out the current locus diagram as shown in Fig. 99, curve II. If the phase advancer is unsaturated, the curve is a circle which passes through the no-load point A of the ordinary circle diagram of the induction motor (curve I) but whose centre is raised above the axis. It is seen that the power factor falls off rapidly as the load is reduced, and at no-load is the same as that of the uncompensated induction motor. In order to overcome this disadvantage to some extent, it is usual to work with very high saturation in the phase advancer at full-load current so that the correcting voltage obtained at light loads is greater in proportion. The current locus curve then takes the shape shown in Fig. 99, curve III, and it is possible in this way to maintain, for example, approximately unity power factor between full-load and a third of full-load. At no-load, however, the wattless current is still practically the same as that of the induction motor without a phase advancer.

Since, when the power factor is corrected, the secondary current of the induction motor is not in phase with the induced voltage, the injected voltage, which is at right angles to the current, has a component in phase with the induced voltage. This component, together with the resistance drop in the phase advancer, has the effect of increasing the slip of the motor compared with its normal slip as an induction motor. Fig. 100 shows the vector diagram of

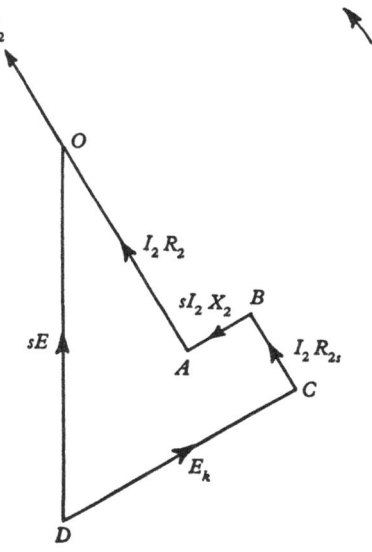

Fig. 100. Vector diagram of secondary voltages in induction motor with Leblanc exciter

secondary voltages, where AO and BA are the ohmic drop $I_2 R_2$ and the reactance drop $sI_2 X_2$ in the motor secondary, CB the ohmic drop $I_2 R_{2s}$ in the phase advancer, DC the injected voltage E_k, and DO the induced voltage sE in the motor secondary. Because of the component of injected voltage which produces a change of speed, the value of injected voltage has to be greater than the minimum required to produce the same amount of power-factor correction. Moreover, the value of the voltage can only be adjusted by varying the speed, a method which is not usually practicable.

The Leblanc exciter thus has several defects. No power-factor correction is obtained at no-load, although the injected voltage is greater than necessary at full-load, and no simple control is possible. The most serious difficulty, however, with this type of machine is the commutation. The use of interpole windings robs the machine of its simplicity, but unless they are used, good commutation depends entirely on the brush, and the effect of high saturation is to introduce high-frequency pulsations owing to harmonics. The most important factor, however, is the fact that the transformer voltage of commutation is always in the same direction as the reactance voltage, because the transformer voltage is induced by the armature flux itself, instead of by an opposing interpole flux. It is, however, possible by careful and liberal design to obtain satisfactory operation with this type of machine.

40. The series-excited phase advancer

In order to overcome the commutation difficulties experienced with the simple Leblanc exciter, the next development was the provision of a compensating winding on the stator to neutralize the armature M.M.F., the main flux being set up by means of a separate series-exciting winding. The phase advancer is then a series-excited Scherbius machine, except that, owing to the low voltage required, it is not always necessary to use the salient pole construction. In order to make the transformer voltage oppose the reactance voltage, the exciting winding is placed so that its M.M.F. is in the same direction as that of the compensating winding; in practice the two windings are generally combined together into one. The commutating conditions are then very good, particularly if the machine is not highly saturated.

In the early machines, the exciting winding was placed so that the induced voltage was in quadrature with the current. The characteristics of the associated induction motor are then exactly the same as those obtained with the simple Leblanc exciter, except that the conditions for commutation are improved.

However, the more usual requirement is for the motor to operate at a high power factor over the whole working range of load, including no-load. Power-factor correction at no-load is possible with a series phase advancer, if the phase angle between the current and the induced voltage is reduced to less than 90° by displacing the exciting winding. The 'Miles Walker' phase advancer is a machine of this type, in which the angle between voltage and current is 30°.

Fig. 101 is a vector diagram of secondary voltages when the motor is on no-load. The lettering corresponds to that of Fig. 100, but now the angle between the current and the injected voltage E_k is less than a right angle.

A closed diagram can be obtained at no-load because, although the current is at right angles to the secondary induced voltage, the injected voltage E_k makes an acute angle with the current. Hence, power-factor correction can be obtained at no-load, as shown in Fig. 102. The actual values of the current

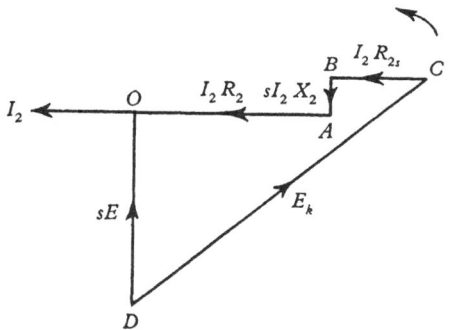

Fig. 101. Vector diagram of secondary voltages in induction motor with series-excited phase advancer

and of the voltage E_k are determined by the saturation curve of the phase advancer, since the ratio of the total resistance drop $AO + CB$ (see Fig. 101) to the voltage DC is equal to the cosine of the angle between them. In other words, the machine self-excites in a similar manner to that described on p. 111.

A good power-factor characteristic, such as that shown in Fig. 102, curve II, can be obtained with this type of machine, because the value of the injected voltage, which is determined by saturation, remains approximately constant over a wide range of load. In order to self-excite with certainty at no-load, the machine must work with high saturation. But the most serious defect in earlier machines was a tendency to hunting troubles, which are liable to occur if the direction of rotation of the field is opposite to that of the armature. When hunting occurs, the phase advancer self-excites at a different frequency from that of the normal secondary voltage of the motor. The result is that hunting of the supply current occurs as explained on p. 121.

There is much less tendency for hunting to occur if the rotation of the flux is in the same direction as the rotation of the armature as in a simple Leblanc exciter. This arrangement is used in a modified form of the series phase advancer known as the 'Heyland exciter'. The main flux in this machine acts in the same direction as the armature M.M.F., and a special distribution of the stator winding is necessary in order to compensate the commutation voltages. This machine operates in exactly the same way as the

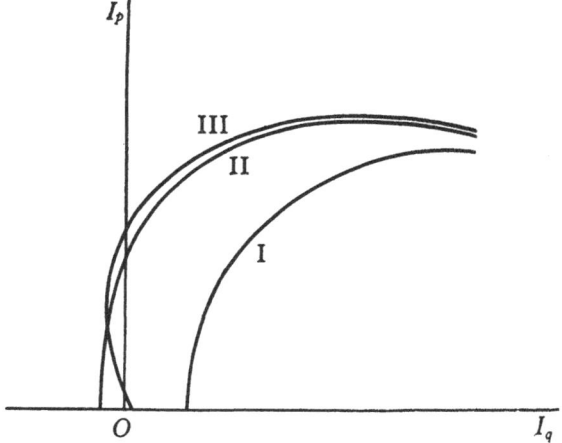

Fig. 102. Current locus curves of induction motor with series and shunt-excited phase advancers

'Miles Walker' phase advancer described above, but, by the special arrangement of the stator windings, hunting troubles are avoided and at the same time good conditions for commutation are obtained.

Adjustment of the power factor of an induction motor with a series phase advancer can be obtained either by varying the speed of the phase advancer, or by the more practical method of diverting some of the current by connecting a resistance across its armature. The amount of adjustment at no-load is determined by the range of voltage obtainable above the bend of the saturation curve.

41. The shunt-excited phase advancer

The shunt phase advancer, like the series-excited phase advancer, is essentially a Scherbius machine, with or without salient poles, but, in the shunt machine, the exciting winding is connected in shunt across the terminals, and the connexions are the same as those of the low-frequency generator illustrated in Fig. 61. The axis of the exciting winding is usually placed 90 electrical degrees away from the axis of the armature, so that the induced voltage in the armature is in phase with the exciting current. The field winding and brushes are then in the same relative position as in a D.C. machine, and consequently the shunt phase advancer, when on open-circuit, will self-excite at zero frequency, as explained on p. 112.

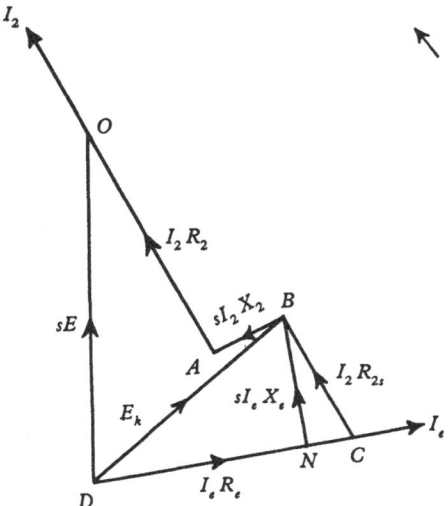

Fig. 103. Secondary voltages in shunt-excited phase advancer

When the machine is connected to the secondary of an induction motor, the frequency of the voltage must be the same as the slip frequency of the motor, and the magnitude, phase and frequency of the voltage generated in the phase advancer are then determined by the fact that the vector diagrams of voltages in the main and shunt circuits must close. Fig. 103 is a vector diagram of the voltages in the motor secondary winding and the phase advancer, all voltages being phase values, assuming an equivalent star connexion for the armature. DO is the induced voltage sE in the motor secondary, AO and BA the resistance and reactance drops I_2R_2 and sI_2X_2 in the motor secondary, and DB the voltage E_k at the terminals of the phase advancer. In the phase advancer, E_k is equal to the vector sum of the induced voltage $DC = E_{kt}$ in the phase advancer and the resistance drop $CB = I_2 R_{2s}$ due to the current in the main windings and in the brushes. The phase advancer reactance is neglected, and as on p. 180, the current in R_{2s} is taken to be I_2.

If the exciting winding is located so that the induced voltage is in phase with the exciting current, the exciting current I_e and hence the resistance drop $DN = I_e R_e$ in the exciting winding and rheostat are in phase with DC. The terminal voltage E_k is equal to the sum of the resistance drop DN and the voltage $NB = sI_e X_e$ which is induced in the exciting winding by the main flux. The foregoing assumes that the armature is exactly neutralized by the compensating winding, that is to say, that there is no flux due to the main current.

This type of machine generates a voltage of approximately constant magnitude over a range of loads in the same way as a shunt-excited D.C. generator. The phase of the voltage depends on the way in which the vector diagram closes, the most important factor being the voltage NB induced in the exciting winding. If the number of turns in this winding is correctly chosen, the phase of the main induced voltage of the phase advancer remains approximately constant over a range of loads, and power-factor correction of the main induction motor is obtained right down to no-load. The current locus curve for a motor with a phase advancer with pure shunt-excitation is similar to that shown in Fig. 102, curve II. In this diagram, curve I is the circle diagram of the uncompensated induction motor. Adjustment of the shunt-field rheostat causes a change in the value of correcting voltage, and moves the curve II nearer to or farther from curve I, thus affording a simple means of adjusting the amount of power-factor correction.

The characteristics may be modified either by changing the location of the exciting winding or by introducing some series-excitation in addition to the shunt-excitation. A particularly simple and satisfactory arrangement is obtained by constructing the machine without salient poles but with a distributed compensating winding having a few more ampere-turns than the armature. This introduces a series component of excitation which compounds the power-factor characteristic by increasing the correcting voltage as the load increases, if the rotation of the armature is opposite to that of the field. It neutralizes the effect of the motor reactance and modifies the current locus curve in the manner indicated by curve III of Fig. 102, in which the wattless current is approximately constant over a wide range of load.

In the shunt phase advancer with over-compensation, the transformer voltage induced in the short-circuited coil due to the shunt and series-excitation, opposes the reactance voltage; there is, moreover, no need to work with high saturation. Consequently the conditions for commutation are very good in this type of machine.

42. The Osnos motor

The commutator frequency changer may be used as a phase advancer, and is well suited to provide a constant injected voltage and to maintain the power factor of an induction motor at a high value at all loads. Nevertheless, it has only found a limited application as a phase advancer, mainly because it has to be direct coupled to the induction motor. The same principle of operation is used, however, in the Osnos compensated induction motor, which has been used extensively for drives where a motor is required to run at unity or leading power factor. The characteristics of an induction motor with frequency changer phase advancer are the same as those of the Osnos motor.

The Osnos motor is like the Schrage motor except that the voltage provided by the commutator winding is much smaller, and no provision need be made for varying its magnitude. The motor has only a single set of brushes, since adjustment of the power factor can be obtained by shifting the brushes so as to change the phase of the injected voltage. In Fig. 104, which is a diagram of connexions of the Osnos motor, the secondary winding on the stator is connected at one end to the commutator, and at the other end to a starting resistance, which is short-circuited when the machine is up to speed.

The characteristics of the Osnos motor are similar to those of an induction motor with a constant injected voltage, when the in-phase component of injected voltage is small. It is clear that the theory of the Osnos motor is exactly that of the Schrage motor if the angle θ is taken to be constant, but the performance obtained is very different because the regulation ratio is only a few per cent, and the principal component of injected voltage is the quadrature component E_{kq}.

If the brushes are in the neutral position, defined in the same way as for the Schrage motor, the motor runs light at a speed slightly below synchronous speed, and no power-factor correction

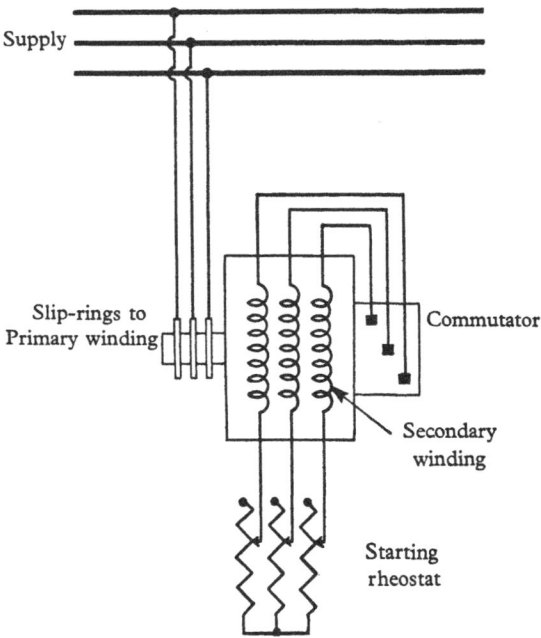

Fig. 104. Connexion diagram of Osnos motor

is obtained. To correct the power factor, the brushes must be shifted backwards away from neutral. As the angle of shift is increased, the power-factor correction increases and the speed rises. With 90° shift, the maximum correction is obtained and the no-load speed is synchronous speed. With an angle of shift greater than 90°, the speed rises above synchronism and the correction falls off, until at 180° no correction is obtained. The neutral position is sometimes called the 'sub-synchronous neutral' and the position 180° away the 'super-synchronous neutral'.

Fig. 105 shows some typical current locus curves for the Osnos motor. The circles have their centres much closer to the horizontal axis than those obtained with the Schrage motor. When the angle of shift is less than 90° the centre of the circle is below the axis, and when the angle of shift is greater than 90° the centre is above the axis. The latter arrangement is often preferable because the wattless component remains more nearly constant over the working

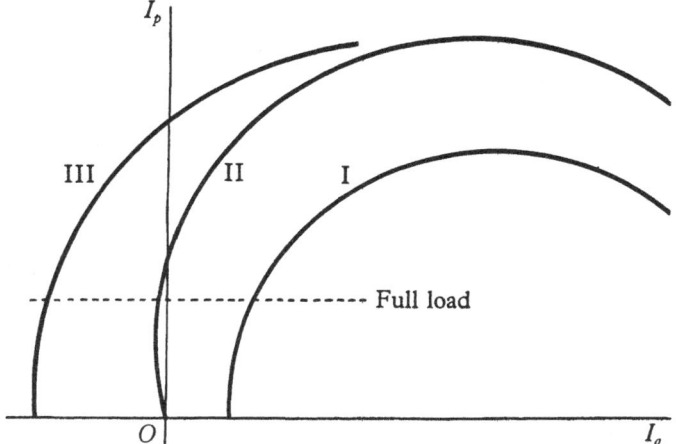

Fig. 105. Current locus curves of the Osnos motor

range. In Fig. 105, curve I is the current locus curve when the brushes are on neutral, curve II corresponds to an angle of shift between 90° and 180° for which the motor operates at unity power factor, and curve III corresponds to an angle nearer to 90° for which the power factor is leading. The power-factor characteristics of an Osnos motor are very satisfactory, as the wattless current is almost independent of the load. Moreover, the pull-out torque is increased as a result of the power-factor correction.

In the Osnos motor the amount of kVA. which has to be carried by the commutator and its winding is only a very small proportion of the kVA. of the motor. The size of the commutator is determined by the size of the motor shaft, and hence is always ample for its duty. Consequently the conditions for commutation are very easy.

Chapter 10

APPLICATIONS OF POLYPHASE COMMUTATOR MACHINES

43. Field of application of variable-speed motors

In view of the general predominance of A.C. supplies for the generation, transmission and distribution of electric power, it is fortunate for engineers that most electric motor drives either require or can tolerate constant or nearly constant speed; it enables them to make wide use of the induction motor which is supreme for simplicity, robustness and cheapness.

But in spite of its large undisputed field of application and well-deserved popularity, the limitations imposed by its 'nearly constant' speed characteristic and its lagging power factor sometimes make the induction motor unsuitable. It then becomes necessary to consider other types of machine, among which polyphase commutator motors or induction motors with regulating machines have an important place.

The expression 'variable speed' is not always used with precision, since it may indicate speed adjustment in large steps, or a continuous variation. The latter is inaptly described as an 'infinitely variable speed' from the notion of an infinite number of steps. For some duties speed control in a small number of steps is sufficient, but for other drives a continuous control is essential.

There is no single generally accepted method of providing a variable-speed drive when the power supply is an alternating current system. The number of available methods is large and many factors must be considered in making a choice between them. Sometimes the conditions clearly favour one method, but at other times all the possible methods have advantages and disadvantages which must be carefully weighed. It is helpful to make a brief survey of the available methods of speed control.

Separate mechanical and other devices used with constant-speed motors

Change-speed gears or stepped pulleys provide adjustment of the speed in steps, while various types of 'infinitely variable' gears, or cone pulleys, give a continuous adjustment. Hydraulic motors, supplied from pumps driven at constant speed, are also used for small powers. These types of apparatus, although still in wide use, tend to be superseded by electrical methods.

For larger powers, particularly for the driving of fans and centrifugal pumps, hydraulic couplings, in which the slip-power is wasted as heat, are frequently used. A similar result is obtained with eddy current or magnetic couplings.

Conversion to direct current

Other methods of obtaining variable speed involve the conversion of the power to direct current and the use of a D.C. motor. In a Ward Leonard equipment the conversion is by a motor-generator set, and the speed of the D.C. motor driving the load is controlled by varying the D.C. generator field. This is the most flexible of all variable speed equipments, but it requires the use of three machines of full capacity with consequent high cost and heavy losses. Less expensive and more efficient is the use of an electronic rectifier instead of the motor-generator set; the motor speed is then varied either by field control of the motor or by grid control of the rectifier so as to vary its output voltage.

Rotor control of induction motors

Speed control of a slip-ring induction motor by means of a variable secondary resistance is in general use for starting up, and is frequently used for intermittent or continuous running below full speed. The slip-power is wasted in the resistance and the efficiency is low, but this is not important for intermittent duties such as the driving of lifts and cranes. For steady drives the fact that the speed varies with the load is often a serious disadvantage.

There are two other methods, not involving the use of polyphase commutator machines, of varying the speed of an induction motor by applying a voltage to the rotor circuit. In a Kramer equipment, the slip-power is passed on to a rotary converter, which runs at a speed corresponding to the slip frequency, and supplies a D.C. motor. The D.C. motor is mechanically coupled either to the main induction motor or to an auxiliary machine, and adjustment of its field controls the speed of the main motor. The second method is the use of another induction motor, cascaded with the main motor and mechanically coupled to it. This enables one additional speed to be obtained.

Change-pole induction motors

By means of special windings and suitable control gear, several speeds—usually not more than four—can be obtained with an induction motor. This type of motor is used extensively for small powers, particularly for driving machine tools. Larger change-pole motors, often with only two speeds, are used in conjunction with hydraulic couplings or rotor resistance, or as part of a cascade set in order to extend the range of variation.

Induction motors with speed-regulating machines, and commutator motors

In general, commutator motors are used for small and medium powers, and induction motors with regulating machines for large powers. Most types of motor or equipment provide continuous speed variation, but some systems of control using Scherbius machines give variation of speed in steps depending on the taps of a transformer. The characteristics obtained may be of the shunt type, the series type, or the drooping type required for fly-wheel drives.

Much the commonest requirement is for a variable speed motor with shunt characteristics, that is, a motor whose speed, when once set by an independent controlling device, remains approximately constant in spite of changes of load. This requirement is met by a Schrage motor, a shunt motor, or by a Scherbius equipment using a regulating machine with shunt or separate excitation.

The Schrage motor is a self-contained motor, normally requiring no additional apparatus except the supply switch, and is a very useful motor for small and medium powers. The output per pole for which the Schrage motor can be made is limited, and it cannot be built for direct connexion to a high voltage supply, but within these limits, good shunt characteristics as well as high efficiency and power factor are obtainable. A wide speed range extending right down to standstill is readily obtained, and is in fact normal in small motors, but in larger motors the possible output at a given speed is somewhat reduced if a long speed range is required.

The shunt motor is less limited in respect to its output and supply voltage than the Schrage motor, particularly when a long speed range is not required. At a given speed a higher output can be obtained than from a Schrage motor, and moreover, the motor and regulator can be supplied at high voltage. The torque characteristics, efficiency, and power factor are all comparable to those of a Schrage motor in the upper part of the speed range, but they are inferior in the lower part, particularly when a long range is provided.

The series motor can often be used for steady drives, but its series characteristics are generally a disadvantage, because the speed is sensitive both to load changes and to voltage changes. The limiting horse-power for which a series motor can be built at a given speed is similar to that of the shunt motor, and is higher than that of a Schrage motor.

Remote control and automatic control

The location of the control device of a variable speed motor is often important. When it is inconvenient for the operator to move the regulator, brush spindle, or other control mechanism directly, the movement can be effected by a small pilot motor controlled from one or more push-button stations. Such a station has push-buttons for 'Start', 'Stop', 'Raise Speed', 'Lower Speed', and may include other controls.

In other installations, the starting and stopping or the speed may be controlled automatically in dependence on the resulting operation of the driven apparatus. Many types of commutator motor, and particularly the Schrage motor, are well suited to this method of control, because of the stepless variation of the speed, and the lightness of the operating gear. Several examples of duties requiring remote and automatic control are described in this chapter.

Choice of drive

There is inevitably a good deal of overlapping in the application of different types of variable speed drive, and the correct choice is not always obvious. The final selection must often be made on a comparison of the cost of the alternatives. It is not easy to generalize, and each case must be considered on its merits, as to the choice between commutator machines and other methods of control or between the different types of commutator machine.

44. General considerations relating to power-factor correction

The advantages of maintaining high power factor on A.C. supply systems are well known. In order to pass on equitably the cost of maintaining the supply to the consumer, any tariff must be related in some way to kVA. demand as well as to the kWh. consumed, because the kVA. demand directly affects capital charges. In some installations, improvement of the power factor may make possible increased utilization of existing cables or transformers. A good power factor is also important from the point of view of voltage regulation.

The main source of the lagging current in a system is the induction motor load, and the high popularity of this motor often makes the problem acute. Overall power factor can often be considerably improved without the use of correcting apparatus by careful examination of the operating conditions of each induction motor. Many motors are left running light for prolonged periods where they could be shut down; another common fault is 'overmotoring', and measurement of the actual currents drawn by individual motors often leads to the use of smaller motors and a higher operating power factor.

Frequently, however, it is profitable to bring up the power factor of an installation to a higher value than is possible by this means, by using types of motor which operate at high power factor, or by using static capacitors. The former method is preferable when the motors are of sufficient capacity and when they run for sufficiently long periods, as it has the advantage of attacking the problem at its source. There is a double gain; first, the omission of the lagging current of the induction motor; and secondly, the injection of wattless current into the system.

Comparison of types

The types of apparatus used for improving power factor may be classified as follows:

(1) Synchronous motors.

(2) Induction motors with phase advancers, or compensated motors.

(3) Commutator machines used primarily for speed variation.

(4) Static capacitors.

The importance of commutator machines for power-factor correction has become less in recent years owing to developments in synchronous motors and static capacitors. For large powers a synchronous motor is generally more efficient than an induction motor with a phase advancer, although in special circumstances a phase advancer may still be appropriate—for example, where a large induction motor already exists. For drives of medium power there is a definite field for such machines as the Osnos motor, which provides power-factor correction at the source, in a simple machine, with more flexible characteristics than those of a bank of capacitors. It is not economical to use this motor for powers below about 20 H.P. or for those larger drives where the motor is shut down for long periods. For the large number of installations consisting entirely or almost entirely of small motors (20 H.P. and below) the most satisfactory method of power-factor correction from every point of view is the installation of static capacitors.

When for any reason variable speed commutator machines are used, a considerable amount of valuable power-factor correction can sometimes be obtained as an incidental advantage. This point should be borne in mind when variable speed drives are under consideration.

Whatever apparatus is used for power-factor correction, the effect under any particular condition is to introduce a definite amount of wattless kVA. into the system. Static capacitors provide at all times constant wattless kVA., which can only be varied by switching sections in or out. When machines are used, the wattless kVA. introduced depends on the load or other operating conditions. Hence the current locus curve of the machine may be important. If the load varies, it is an advantage to use a motor of a type with which the power-factor correction is obtainable at all loads.

45. Power stations and power systems

Auxiliary drives in power stations

In the design of generating stations, the choice of auxiliary apparatus is rightly regarded as an important matter. The performance of the station depends to a considerable degree on the design of the auxiliary services, with the result that continual efforts are being made towards more efficient and more flexible drives. This has inspired the study of variable speed control of auxiliaries, and polyphase commutator motors have accordingly been installed in many stations, with noteworthy success.

Boiler-house auxiliaries

The most important application of variable-speed motors in a power station is for auxiliary drives in the boiler house, whether the fuel is in solid or pulverized form. The varying conditions of load and grades of fuel call for flexible means of control both of the supply of the fuel and of the air required for combustion, and automatic control in dependence on the output is often used. The ash handling plant also may be driven by variable-speed motors.

The forced draught and induced draught fans require fairly large motors, often over 100 H.P. The methods which have been used for controlling the supply of air are:

Damper control (baffling of the fan output) with squirrel-cage motors.

Vane control (varying the angle of the stationary vanes) with squirrel-cage motors.

Vane control with change-pole squirrel-cage motors.

Hydraulic or electric slip couplings with squirrel-cage motors.

Rotor resistance control with slip-ring motors.

Direct current motors with conversion plant.

Polyphase commutator motors of several types.

The first two methods, where the fan runs at a constant speed, are very inefficient, but vane control is often used in conjunction with a two-speed motor. Apart from the question of efficiency, reduction of the fan speed lessens the wear on the fan system, especially where the gases contain abrasive ash; with variable speed the fan rarely operates at maximum speed and there is therefore a large increase in the life of the fan. Fig. 106 shows typical efficiency curves for several of these methods. The advantages of commutator motors are obvious.

Similar considerations apply to the drive of the forced draught fans, and of the exhauster fans in pulverized fuel stations.

The choice of the method of obtaining the variable speed is mainly an economic one. The need to enclose the motor so as to exclude the dust of the boiler house is an important factor. The motors, which are too large for ordinary total enclosure, may be ventilated through filters, but the filters need a great deal of attention, and it is more satisfactory to use a closed air circuit for the air in the machine, in conjunction with a heat exchanger. Either shunt or series commutator motors or Schrage motors may be used for these drives, but the shunt motor is preferable for this

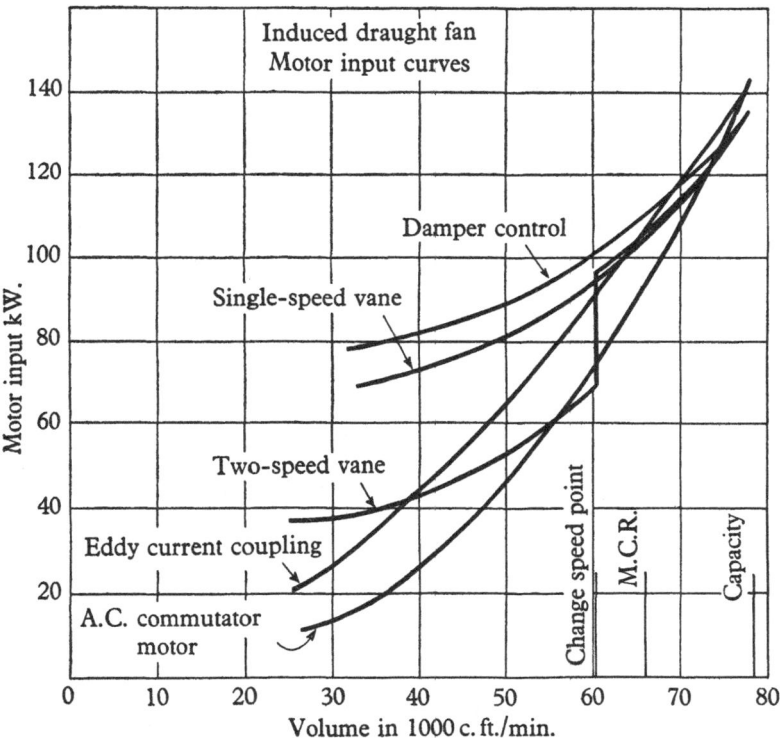

Fig. 106. Efficiency of various drives of induced draught fan

duty, because the enclosed ventilation is easier to arrange for a motor with fixed brush gear, and because it is often an advantage to use a high voltage supply. Automatic combustion control can readily be incorporated with either type of motor.

Chain-grate stokers and ash handling plant, or the feed mechanism in a pulverized fuel station, are often driven by variable-speed motors. Fig. 108* shows a chain-grate stoker driven by a Schrage motor. These small drives require a constant torque

* Figs. 108–125 are at the end of the book.

with varying speed, or even increased torque at low speed, in contrast to the fans, for which the torque drops rapidly with the speed. The motors are totally enclosed or totally enclosed fan cooled. Both Schrage motors and shunt motors have frequently been used for such drives, often in conjunction with automatic combustion control. Where adjacent grates are required to move at the same speed they may be driven by two shunt motors controlled by a single regulator.

Other auxiliary drives

Circulating water pumps and boiler feed pumps are generally driven by constant-speed induction motors. Special conditions, however, sometimes call for variable speed and several installations using commutator machines have been made. Examples are:

(a) Circulating water pumps which have to cater for varying heads of the water supply.

(b) Boiler feed pumps for very high steam pressures.

For such drives, where the pumps are of the centrifugal type, the output is usually large and the speed range small, and some form of Scherbius control provides a solution.

Fig. 109 shows two single-range Scherbius equipments driving circulating pumps. Fig. 110 shows an induction motor with a gear-driven frequency changer, which forms part of a Scherbius equipment with separate excitation. The equipment drives a high-pressure boiler feed pump.

A special control for 'phase changing' pumps for boiler feed is worthy of mention. Here a frequency changer is used to vary the phase relationship between two reciprocating pumps driven by two induction motors, so as to control the quantity of water. Fig. 111 shows the frequency changer, the commutator brush gear of which is connected to the secondaries of the two induction motors, causing them to run in exact synchronism with each other (see p. 117), but at a space phase angle which can be varied.

Application of commutator machines for system control

Apart from their use for driving auxiliaries in the generating station, commutator machines have been applied in two notabl ways for controlling the external power system:

(a) Variable-ratio frequency converter sets.

(b) Asynchronous condensers.

These applications, which often require very large units, use an induction motor with a regulating machine, generally coupled together on the same shaft. Two alternative types of regulating machine have been used:

(a) The Scherbius machine with separate excitation from a small frequency changer exciter (see Section 35).

(b) The commutator frequency changer, generally of the compensated type.

Many variations in the details of the control are possible.

Variable-ratio frequency converter set

The simplest and most commonly used type of frequency converter set for interconnecting two power systems of different frequency consists of two synchronous machines, one connected to each system. With this arrangement, however, the flow of power cannot be controlled from the set itself, because of the fixed ratio of frequencies in the synchronous machines. Where both supply systems are of large capacity, particularly if one system supplies a traction load, it becomes necessary to introduce some means of varying the ratio of the frequencies. The most flexible means of providing this variation is to make one unit of the set an asynchronous machine with a commutator type regulating machine.

In one important type of variable-ratio converter set, one unit is a synchronous machine with a D.C. exciter, and the other unit is an asynchronous machine with a Scherbius type regulating machine and a commutator frequency changer with movable brushes. The connexions of the asynchronous part of the equipment are those shown in Fig. 90, and the operation is explained in Section 35. To a first approximation, displacement of the brush gear by means of a cam-operated mechanism, controls the relation between the slip of the asynchronous machine and the load, and variation of the resistance in the auxiliary exciting circuit controls the wattless current. Each control is operated by a pilot motor, in conjunction with wattmeter relays or other devices, and complete control of both real and wattless power is obtained for any ratio of frequency within the operating range for which the equipment is designed.

The range of operation of the asynchronous machine generally covers both motoring and generating, and both leading and lagging conditions, with variations of frequency and of voltage. The usual requirement is that the power transmitted shall be held constant automatically for any setting of the manual control in spite of variation of the frequency of either system, but it is easy to cater for other requirements for special purposes.

An alternative method of control for a Scherbius machine uses a frequency changer exciter with fixed brushes, the voltage of which is varied in both magnitude and phase by means of double induction regulators or special synchronous machines. This provides the same independent control of real power and wattless power. When the main regulating machine is a compensated frequency changer, a similar method of control by means of induction regulators is used, but the power handled by the control apparatus is reduced. Additional apparatus, such as special transformers or machines, are sometimes included in the excitation circuits for special purposes, but the general principles involved in the automatic control are the same whatever the details of the system.

Asynchronous condensers

The use of large synchronous machines as condensers for the control of the wattless current in power systems is well known. The asynchronous condenser, which consists of an asynchronous machine with a regulating machine for controlling the wattless current, serves the same purpose as the synchronous condenser. The advantage claimed is that the asynchronous machine is more stable during system disturbances, particularly when supplying reactive kVA., but it is considerably more costly, and in view of recent improvements in the design and application of synchronous machines, it is doubtful whether any more equipments of this kind will be made.

46. Paper-making machinery

Paper-making has become an important industry in the modern world; the quantity of paper required not only for book publishing, newsprint and general administration purposes, but for industry in general, has vastly increased in recent years. Few industries have made such rapid strides; this expansion has been accompanied by a change-over from steam to electric drive, and the electric drive has proved to be so convenient and so efficient that it is invariably adopted for new mills, and the conversion of old mills is constantly occurring. Even when electric power is not readily available, the electric drive is so advantageous as to make the installation of generating plant worth while.

Constant-speed drives

Many drives in the paper mill require large or medium sized constant-speed motors. Pulp grinders may require several thousand horse-power. The pulp from the grinders is cleaned and boiled with solvents, and then broken up into its constituent fibres in breaking machines such as beaters, potchers and refiners, which prepare it for the main paper-making machine. The motors for the preliminary processing machinery are constant-speed machines requiring 100–500 H.P.

Any or all of the larger constant-speed motors, which are in operation day and night for weeks at a time may be high power factor machines. Induction motors with phase advancers and Osnos motors have frequently been used for these drives, but synchronous motors are nowadays used to an increasing extent because of their higher efficiency, and recent improvements in starting methods.

Variable-speed drives

The main paper-making machine and the machines for many of the succeeding processes operate at variable speed. The main machine consists of several sections, each performing a separate operation, while the paper passes through each in succession. The whole machine, particularly when its size is small, may be driven by a single variable-speed motor, and provision must then be made to adjust by mechanical means the relative speeds of the various sections. But for larger machines, considerable economy and greater ease of control is obtained by providing a separate motor for each section, and regulating the speed of each section by means of special automatic controlling apparatus. The sectional drive for paper machines is an important application of A.C. commutator motors, and is described in more detail on p. 208.

The processes following the main paper-making machine also often need variable speed. The super-calender consists of a vertical stack of rolls, alternately of highly polished steel and of composite material. The paper is uncoiled from a winder, passed through the rolls and recoiled on a rewinder. This operation gives the paper surface a fine smooth finish. When different weights and grades of paper have to be handled, a wide range of variable speed is necessary, for which either the Schrage motor or the shunt motor may be used. A large super-calender driven by a Schrage motor is illustrated in Fig. 112.

A creeping speed must be provided for threading the paper into the calender, an operation which takes place as often as three or four times an hour, whenever the rolls are changed. It is usual to provide a barring motor consisting of an induction motor geared down to the creeping speed. The barring motor also gives the desired high starting torque. It is used to start the super-calender, and holds it at creeping speed until the operator is ready to speed the machine up to calendering speed. He then presses the 'Raise Speed' button and this switches on the commutator motor which takes over the load from the barring motor and continues to increase speed until the button is released.

Similar equipment is employed for the winder and rewinder drives but on a smaller scale. Once the calendering speed is attained, the super-calender motor runs at constant speed, but the winder and rewinder motors must have a gradual speed change as the diameter of the roll increases or decreases. They usually require a wide speed range extending down to zero speed.

In addition to the motors concerned in actual paper manufacture, there are auxiliaries in the mill such as lifts and cranes for which variable speeds are necessary and polyphase commutator motors are suitable.

Sectional drive for paper-making machine

The equipment used for controlling independently the motors driving the sections of a large paper machine is an outstanding example of the automatic control of electric motors. Direct current motors can be used for this purpose, and at least one installation has been built using shunt motors. The Schrage motor has the advantage over the shunt motor that the control mechanism is lighter and that the torque-speed characteristics are superior at low speeds.

The first section consists of the Couch rolls, and it is followed by a number of Press roll sections, Drying cylinder sections,

Cooling cylinders, and finally the Calender and Reel sections. It is very important that there shall be fine and stable speed control, because even small inadvertent speed changes cause variations in paper thickness and non-uniformity of texture. In the sectional drive the relative speeds of the various sections must be correctly set and maintained.

Between the several sections it is necessary to have a difference, known as the 'draw', in the peripheral speed of the rolls. As the paper is in the process of manufacture there is a continual quickening of the paper speed from the initial Couch section up

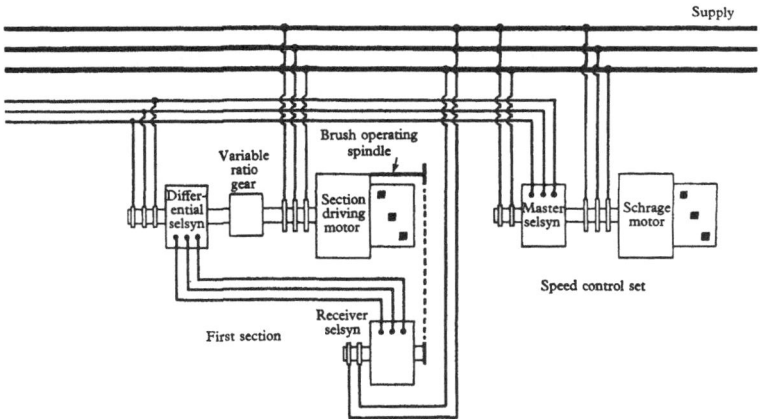

Fig. 107. A.C. sectional drive of paper machine
with selsyn control

to the Reeler. The correct amount of draw has to be determined by experience and, once set, has to be held constant or the paper will break; the sheet is exceedingly weak, especially at the wet end. Hence there must be an accurate automatic regulation of the speed of each motor, combined with a fine control for adjusting the draw between successive sections.

Selsyn regulating system

The connexions for a typical A.C. sectional drive using Schrage motors and selsyns are shown in Fig. 107. A selsyn is a small electrical machine the speed of which is strictly proportional to the difference between the frequencies applied to its stator and rotor windings.

The speed of each Schrage motor driving a section is controlled by a selsyn system which matches it against a master speed. This master speed for the whole paper machine is determined by a separate speed control set consisting of a master selsyn driven by another Schrage motor.

The control system for one section driving motor consists of the master selsyn, a differential selsyn driven from the motor through a variable ratio gear, and a receiver selsyn coupled to the motor brush gear. The stator of the master selsyn is connected to the rotor of the differential selsyn, the stator of which supplies the receiver selsyn. In a condition of equilibrium, when the section motor runs at exactly the right speed, the speed of the differential selsyn agrees with that of the master selsyn, and the voltage applied to the receiver selsyn holds that machine at rest. If, however, the speed of the section motor varies, the voltage applied to the receiver selsyn varies so as to cause it to rotate the motor brush gear and restore the speed to the correct value. In practice, the speed correction is so rapid that any tendency for the speed to wander is immediately checked.

While the speed control set runs at constant speed, all the section driving motors must also run at constant speed. If the speed of the set is raised, all the section motor speeds rise in proportion to conform with the change in speed of the master selsyn. The speed of the whole paper machine is thus controlled by the brush separation of the Schrage motor driving the master selsyn. This is effected by a pilot motor with remote control usually from a push-button station at the front of the paper machine.

The differences in speed between the sections necessary to allow for draw of the paper, are obtained by means of the variable ratio gears between the motors and the differential selsyns. A very fine control of the draw is obtained at all points by adjusting the variable ratio gears.

Synchronous motor regulating system

An earlier version of the A.C. sectional drive using the Schrage motor also works on the speed-matching principle. The brush operating spindles of the section driving motors are driven by a regulating gear, which incorporates a small synchronous motor of special construction in that the armature is free to rotate as well as the field system. The armature is belt-driven from the shaft of the section driving motor, and when the speed at which it is driven is its synchronous speed, the field member remains stationary. The field member is mechanically connected to the brush-operating spindle of the driving motor; any departure from the synchronous speed of the regulating motor can only be caused by a change in driving motor speed, and this change is promptly corrected by the movement of the field member.

The regulating motor is supplied from the speed control set, which consists of an alternator driven by a Schrage motor, and thus the frequency of the supply to the regulating motor is determined by the speed of the control set. Hence the speed of the whole paper machine is controlled by the brush separation of the Schrage motor of the speed control set, just as in the selsyn scheme.

The belt drive between section motor and regulating motor is by cone pulleys. This allows the speed of any individual section to be adjusted independently, for setting the 'draw'. A guide pulley, remote controlled from the front of the paper machine, determines the position of the belt on the cone pulleys and thus the draw is under the direct control of the operator. Fig. 113 shows a paper-making machine with an a.c. sectional drive using the synchronous motor type of control.

47. Steel and metal industries

Many of the early applications of commutator machines were made in steel mills, particularly for large drives using induction motors with regulating machines. More recent developments have, however, led to a demand for larger powers and longer speed ranges from the motors, and large rolling mill drives now commonly use d.c. motors with Ward Leonard control. There are, however, still some uses for commutator motors, particularly for the smaller mills, and for such duties as strip-winding and wire-drawing.

Scherbius drives for large rolling mills fall into two alternative classes, depending on whether or not a fly-wheel is used.

(a) If the peak loads on the mill are not more than a normal motor can carry, it is not necessary to use a fly-wheel, and the motor is required to have shunt characteristics. Speed variation is needed in order to cater for different sections, some of which can be rolled at higher speeds than others. Many large Scherbius equipments of both the single and double-range type, have been installed in Europe and America for driving rolling mills of this kind.

(b) In sheet mills and other types of mill, where very heavy peak loads are encountered, a fly-wheel is provided, and the motor must have a drooping characteristic in order to allow the fly-wheel to give up some of its stored energy when the peak load occurs. Induction motors with series-excited Scherbius machines have been used for this purpose, sometimes having a single characteristic, and sometimes having an independent control of the no-load speed as well as the drooping characteristic. Sometimes the induction motor driving the generators of a Ward Leonard Ilgner equipment operates in conjunction with a series-excited slip regulating machine.

Fig. 114 shows a 700 H.P. motor driving a sheet mill, where a Scherbius machine with combined series and separate excitation provides a drooping characteristic together with power-factor correction at all loads. The small machine on the left of the motor is the frequency changer.

Several constant-speed mills have been equipped either with induction motors and phase advancers, or with Osnos motors. For smaller variable-speed mills the Schrage motor or the shunt motor provides a convenient method of drive. For wire-drawing machines a small commutator motor of a few horse-power is very suitable.

Because of the ease with which the Schrage motor can be controlled, it has been used for some interesting installations where a fine control is essential. When narrow strips are to be rolled, particularly non-ferrous material such as copper and aluminium sheet, the power required falls well within the range of the Schrage motor. An installation of this type is shown in Fig. 115, where the Schrage motor is remote controlled from a push-button station close to the operator. The brush-shifting unit, consisting of pilot motor and gear, is mounted on the top of the Schrage motor, and immediately below it, is a device for pre-setting the speed. If for any reason the motor is shut down, the pilot motor automatically returns the brush gear to the low speed position so as to be ready for re-starting. The object of the pre-set device is to ensure that when the operator starts the motor again, he can bring it up to exactly the same brush separation and therefore the same speed. On pressing the start button, he starts the motor, and the brush-shifting mechanism raises the speed to that indicated by the pre-set device.

48. Lifts, cranes and mine-winders

Lifts

The electric motor has now displaced nearly all other forms of power for lift service both for goods and passengers. The characteristics demanded from the motor depend on the type and size of lift, the speed required and the duty cycle. High speed gearless lifts driven by D.C. motors are very common in America, but in Europe lifts are generally driven through gears. For geared lifts three types of motor drive are available:

(1) Single- and two-speed induction motors.

(2) Direct current motors with Ward Leonard control.

(3) Schrage motors.

The main operating speed of a lift has a fixed value, but the operation at full speed is less important than the conditions during acceleration and retardation. The motor accelerates from rest, but it must slow down to a definite levelling speed in order to come to rest in the correct position. The acceleration and retardation must be rapid, but must nevertheless be accomplished smoothly, especially in passenger lifts, for it is the rate of change of acceleration which causes discomfort to the passengers. Direct current motors and Schrage motors both give smoother operation than induction motors because a stepless control is obtained.

The Schrage motor drive is more efficient than either of the alternatives. The change-pole induction motor wastes in each trip nearly twice the energy stored in the moving parts in addition to the motor losses, whereas the Schrage motor recovers most of the stored energy by regenerative action. A Ward Leonard D.C. equipment suffers from the disadvantage that the motor-generator set consumes power even when the lift is not moving. This loss, which is considerable, is avoided when a Schrage motor drive is used.

The control gear used with a Schrage motor is very simple. The brush gear is moved by means of a small pilot motor at a rate which determines the acceleration and retardation. Normally the brushes move over the complete travel in about two seconds when accelerating from rest, and one second when slowing down to the levelling speed, but the resulting rapid change of speed is achieved without discomfort, mainly because the speed control is stepless. The control gear consists of two reversing contactors for the main motor, and two small contactors for reversing the pilot motor operating the brush gear. A typical lift installation is shown in Fig. 116.

Any lift motor must operate without making an undue amount of noise, which can be very objectionable in residential and business premises where the majority of passenger lifts are installed. The most objectionable noise in a lift motor is the magnetic hum, which is transmitted through the framework and foundations; windage noise and brush noise are both airborne and therefore largely confined to the lift house.

The magnetic loading of the Schrage motor is necessarily light compared with that of an induction motor of the same output, and therefore the most offensive form of noise is not troublesome. Owing to the intermittent nature of the load, and the brief period of full-speed running, it is unprofitable to fit large ventilating fans, and windage noise is therefore small. Bearing noise is eliminated by using oil-ring lubricated sleeve bearings instead of the more usual ball and roller bearings. The source of noise which requires most care is the commutator. Measures to reduce this noise are always taken: a soft graphitic brush is used, and the slots between commutator segments where the micas are undercut are filled with a soft adhesive and insulating compound. The resulting surface is smooth, and a highly polished finish is obtained. Such measures are quite effective, and since what noise still remains is airborne, it can fairly be said that when lifts are driven by Schrage motors, noise is no longer a problem.

Cranes

Closely allied to the drive of lifts is the hoisting motion of cranes, where a very fine control is desirable. The operation is less simple than that of a lift because a range of operating speeds is required. The speed of a crane must be under the immediate control of the crane driver, and the response to his control handle must be rapid. In a lift installation, the load is partially balanced, while the crane has to handle any load from an empty hook to full load lowering and hoisting, and any operating speed may be wanted from creeping to maximum speed.

It was at one time thought that motors for crane duty should have series characteristics. Further investigation has shown, however, that for most types of crane shunt characteristics are preferable provided that a sufficiently wide speed range is available. The use of D.C. series motors on cranes in preference to shunt motors is often due to the limited speed range of the D.C. shunt motor.

The Schrage motor can have a very wide and stable speed range, and provides an excellent means of driving a crane. Several control schemes have been devised. A control scheme which has given good results uses selsyns, which operate in the manner described on p. 209 in connexion with paper-making machinery. In the application to cranes, the receiver selsyn is coupled to the brush gear of the Schrage motor, while the transmitter selsyn is coupled to the controller handle. The receiver selsyn is supplied from the transmitter selsyn through a differential selsyn, and any difference in position between transmitter and receiver causes the differential selsyn to turn. The movement operates a contact which closes the circuit of the pilot motor driving the brush gear, whereupon the pilot motor either accelerates or retards the Schrage motor until the transmitter and receiver selsyns are again in phase. Thus the position of the brush gear and hence the speed of the Schrage motor are made to follow exactly any movement of the controller handle. Fig. 117 shows a Schrage motor with selsyn control, driving the hoist motion of a crane.

The polyphase series commutator motor has been used for dock cranes, where a very high no-load speed for returning a light hook is desirable. The control is carried out by means of brush movement. A number of additional control devices are often necessary to take care of braking and quick reversal, and to prevent overspeeding, and it is difficult to meet all the requirements without a good deal of complication.

Rating of lift and crane motors

The correct rating of the motors is an important problem in this kind of application. It has been the custom to determine the starting torque required, and then to fix an arbitrary H.P. rating bearing some relation to that torque. On the acceptance tests, the motor is required to be run at this arbitrary rating for ½ hour or 1 hour without exceeding the standard temperature rise; the duration of the heat run depends chiefly on the fancy of the lift or crane manufacturer. The method answers the purpose for small motors of a few horse-power, but its absurdity is exposed when large motors are involved.

Most of the losses in a lift motor are developed during acceleration and retardation, and therefore the mechanical inertia of the system plays a large part in determining what the average losses are. When the lift is in continuous operation in the sense that there is no intermission between trips, it is possible for the actual temperature in service to be excessive, even though the motor satisfied the requirements of the specified tests. Since the greater part of the inertia of the system is that of the motor itself, the replacement of the motor by a larger one can make the temperature in service higher instead of lower, even though the specified test would indicate an improvement. It has, in fact, often been found that when a lift motor operating at too high a temperature has been replaced by a smaller machine, the latter has performed the duty with normal temperature rise. Therefore, when applications with duty cycles like those of lifts or cranes are under consideration, the size of the motors should be chosen to fit the torque requirements, but the temperature rise to be expected should be determined by a study of the duty cycle and the total inertia. If a high temperature rise is expected, it is preferable to employ improved insulation materials suitable for the high temperature. This practice often leads to more efficient operation owing to decrease of inertia. For a temperature test, a service factor based on the duty cycle and the total inertia, instead of on the starting torque requirements, should be introduced.

Mine-winders

A mine-winder is like a large lift, but the conditions are in some respects different. The accelerating and stopping times are a much smaller proportion of the cycle of operation and the resulting losses are therefore less serious. For the smaller winders, induction motors with resistance control provide a satisfactory drive, but shunt commutator motors have been used for this duty.

For large winders, Ward Leonard equipments are more economical and reduce the peak load taken from the supply during
acceleration. The peak load can be still further reduced by using
a so-called Ilgner equipment, in which the Ward Leonard motor
generator set is provided with a heavy fly-wheel, as in the rolling
mill drives described in the preceding section. To bring the
fly-wheel into action the speed of the fly-wheel must drop when
the peak loads are applied. The simplest method is to use an
induction motor with a fixed external rotor resistance or a series-
excited Scherbius regulating machine. More complete and efficient
equalization is obtained if the rotor resistance or the excitation of
the Scherbius machine is automatically varied in dependence on
the induction motor load.

Several interesting Ward Leonard winder equipments have been
installed, using a separately excited Scherbius machine with a
frequency changer exciter having movable brushes, connected as
in Fig. 90. With this method of control, the brush gear of the
exciter is moved by means of a small torque motor, which is
supplied from current and potential transformers in such a way
that its torque is proportional to the power input to the induction
motor. The torque motor is balanced against an adjustable dead
weight. When the power taken by the set exceeds that corresponding to the weight, the weight is lifted and the exciter brushes
are displaced in the direction which causes a reduction in the speed
of the set. Thus energy is released by the fly-wheel so as to reduce
the power taken from the system. Similarly, if the power taken by
the set falls, the weight is allowed to fall by the reduction of torque
exerted by the torque motor, and the reverse process takes place.
Hence the power taken from the supply is maintained constant in
spite of the peak loads on the generator. At the end of each cycle
the fly-wheel regains top speed, and the input falls until the
beginning of the next cycle. Fig. 118 shows the three A.C.
machines and the fly-wheel of a motor-generator set for a large
winder in a gold mine.

With this type of equipment, the power factor of the induction
motor may be corrected so that it operates at approximately unity
power factor under all conditions.

49. Miscellaneous applications

Pumps, compressors and fans

Apart from the pump and fan drives already mentioned in connexion with power-station auxiliaries, many other pumps, compressors and fans are driven by commutator motors or Scherbius equipments when variable speed operation is necessary. The most extensive field for this kind of application has been for pumping water for town supplies or for sewage disposal. In recent years the electric drive has been applied to artesian well pumps, which are now designed for high speeds and high efficiency. A vertical pump of the centrifugal type is used with a vertical spindle motor coupled and mounted directly above it, making a compact unit. In bore-holes giving large yields, seasonal changes occur in the water levels which demand speed variation if high efficiency is to be maintained. Furthermore, it sometimes happens that neither the quantity of water to be pumped nor its pressure head is constant, and it is then desirable to use a variable-speed motor. High efficiency, and often high power factor as well, are most important considerations because the pumps are in operation for prolonged periods at a steady speed and on a steady load. Schrage motors, shunt motors and Scherbius controlled induction motors have all given good results on this duty. A picture of a vertical Schrage motor driving a bore-hole pump is shown in Fig. 119.

Force-pumps for raising the water from the pumping station to the reservoir may be mounted on the same shaft as the well-pump, or if the water requires treatment at the station, the force-pump may be driven separately by another horizontal variable-speed motor. A water supply undertaking may use variable-speed pumps in conjunction with constant-speed pumps, the control being obtained on the variable-speed motors.

Similar conditions exist in sewage disposal installations, except that reciprocating pumps are often used. These large pumps operate at low speed and require a wider range of variation than centrifugal pumps, but good results have been obtained both with Schrage motors and Scherbius equipments. Fig. 120 shows a double-range Scherbius equipment installed in a sewage pumping station. The Scherbius regulating set and the control pillar can be seen in the background.

The pump drive of hydraulic accumulators is an interesting example of the automatic control of commutator motors, and is one of the earliest uses made of the Schrage motor. The brushes are operated by a link mechanism, and the amount of brush separation is made to depend on the position of the accumulator; when it rises the motor speed is lowered and vice versa, with the result that if the equipment is correctly designed the motor is rarely, if ever, shut down, but accommodates its speed to the requirements.

Similar motor characteristics are required for the drive of compressors, either of the centrifugal or reciprocating type, delivering air or gas against a constant pressure. The speed of the motor may be automatically adjusted to suit the volume required.

Large fans for mine ventilation are sometimes driven by Schrage motors or induction motors with Scherbius control. The speed variation is provided to allow for changes in the conditions as the workings in the mine are developed. Blast-furnace blowers have also been driven in this way.

Printing presses

The printing industry is almost as important as the paper-making industry. The electric drive has now superseded all other kinds of printing-press drives, and since accurate control of the speed is an essential feature, polyphase commutator machines have been applied to the duty on a large scale. These motors give that flexibility of control in starting, inching, accelerating to the required speed, and rapid stopping in emergency, which is indispensable for high quality work and high printing speed. In the larger sizes a barring motor is used to obtain inching, and it is frequently necessary to provide a pre-setting control device similar to that described on p. 212, so as to obtain exactly the same speed again after a temporary shut down, because a change in speed may affect the registering of the paper. A small press may be driven by a shunt motor as illustrated in Fig. 121. By locking the regulator, and starting by direct switching, a pre-set control is obtained in a very simple manner.

The printing press imposes a steady load on the motor, and the torque required is approximately the same at different speeds.

Because of this condition, the polyphase series motor has been used successfully for this duty, and has some advantage in smoothness of starting. A steady running speed is obtained in spite of the series characteristics, provided that precautions are taken to ensure stable operation at the lower speeds. Schrage motors and shunt motors are less subject to erratic speed variations due to changes of load or voltage.

Rubber calenders

The motors required for driving rubber calenders are similar to those used for paper calenders. In order to produce consistent sheeting, accurate speed control must be provided; for this reason, Schrage motors are used in many rubber mills. The duty is severe, because prolonged runs at low speeds are quite common. If a shut-down should occur, it may be necessary to back out the partially calendered rubber, and the motor must therefore be reversible. Although varying brush shift can still be used to improve the starting torque and the running characteristics in the forward direction, it must not be carried to the point where the starting torque in the reverse direction is insufficient to back out the material.

Cement kilns

In a cement works most drives are at constant speed. It is only the kiln and the slurry feed to the kiln which may require speed variation. The slurry moves down the rotating kiln by gravity because the kiln is tilted from the horizontal by a small angle. The temperature of the slurry increases throughout its passage down the kiln until the material is fused and leaves the kiln at the lower end as a granulated and partly vitrified solid at a high temperature. The quality of the finished product depends largely on the accuracy of speed of the slurry in this process, and this is obtained by controlling the speed of rotation of the kiln. The drive is difficult, for it involves the rotation at a slow speed of a tube from 8 to 10 feet in diameter, and from 250 to 350 feet long; the kiln is raised to a high temperature by the gases passing up inside. Owing to the mass of the kiln and the weight of the charge, the starting torque is very high, involving the use of a motor capable of such high torque together with some degree of speed variation. Both the Schrage motor and the shunt motor can meet these requirements. The slurry feed motor may be an induction motor supplied from an alternator coupled to the variable speed kiln motor; thus whatever speed changes are made in the kiln motor are passed on to the slurry feed motor.

High power factor motors

There are, as a rule, no special features about applications of phase advancers and compensated motors, because such a machine is used for machinery which could just as well be driven by an uncompensated induction motor or a synchronous motor. Fig. 122, which shows a large induction motor with a belt-driven shunt-phase advancer, is a good example of this kind of installation. The motor drives the machinery of a flour mill.

Low frequency generators

The operation of induction motors at low speeds for starting, inspection or other purposes, can be effected by supplying them from a low frequency polyphase supply. Scherbius machines and commutator frequency changers acting as low frequency generators have been used for this purpose in the paper industry, in sugar refineries and in the steel industry.

Testing equipment

Apparatus for testing is frequently required to operate at variable speed, and there are many interesting applications of commutator machines for this purpose. Small motors may drive variable frequency alternators, or such apparatus as vibration tables or fatigue testers. Fig. 123 shows a variable-speed frequency changer set which supplies a high-speed squirrel-cage motor used for fatigue testing. The regulator controls the speed of the shunt motor, and hence the speed of the specimen being tested.

In the aircraft industry, commutator motors have been used for testing propellers and superchargers, and Scherbius drives have been used for large wind tunnel fans.

General use of small motors

All industries provide many examples of the use of small variable-speed motors. Where possible, single or multi-speed squirrel-cage motors are used because of their simplicity and robustness, and the great majority of small motors are of this type; but there is still a wide field where considerable advantage is obtained if the speed can be varied continuously. Commutator motors are extensively used for such applications.

For most of them there are no special requirements except easy control of the speed by means of a hand-wheel or push-button at an accessible point. Such machines can be standardized and manufactured in quantities regardless of the particular drive for which they will be used. Both the Schrage motor and the shunt motor have suitable characteristics for these duties. The former type is self-contained and is often more convenient to install, but the latter has an advantage if the motor location is inaccessible, because the motor itself is smaller, and the regulator with its hand-wheel can be mounted in any position.

Both types have been used in the confectionery trade for driving travelling conveyors, ovens, and provers, all of which have to deal with a range of products under varying conditions. Standard commutator motors are also used for all kinds of conveyors and drives for continuous processes in other industries.

A few applications of small motors where automatic control or some other special feature is required are described in the following pages.

Ring-spinning frames

In the cotton industry, many thousands of commutator motors have been used to drive ring-spinning frames. The constant-speed drive is still very much used, especially in Lancashire, but in India and on the continent of Europe, a variable-speed motor drive has been generally adopted. With the commutator motor drive, there is an automatic speed regulator designed as an integral part of the motor. This regulator not only reduces the speed at the beginning and end of a spinning period, but also varies the speed periodically as each layer of yarn is built up on the bobbin. The aim is to maintain uniform tension in the yarn, and to give it a constant amount of twist during the building up of the bobbin. The use of automatic variable speed enables a high average speed to be obtained with a consequent large increase in output. Furthermore, it provides a more uniform and a better quality yarn.

Earlier installations of this type used the polyphase series motor because of its simplicity compared with the Schrage motor, but it was later found that serious erratic variations of speed could occur due to voltage fluctuations, and most modern ring-spinning motors are of the Schrage type. Shunt motors have also been used.

The motors are specially designed for the spinning frame, not only to incorporate the automatic speed regulator, but to achieve the minimum possible length. They have to be duct ventilated from a common clean air duct.

Hosiery machines

Knitting machines for fully-fashioned hosiery have unique requirements. The knitting of a complete stocking entails special operations such as knitting plain courses, fashioning, welt turning, plating, clocking, and picoting. These operations are carried out at different driving speeds and impose varying loads on the motor. During the knitting of plain courses, the speed is adjusted to suit the material, and may reach eighty courses a minute. During fashioning, the speed is reduced rapidly and smoothly to about forty courses a minute before each narrowing operation, after which the speed is restored. Experience has shown that A.C. commutator motors are admirably suited to this duty.

When a Schrage motor is used, the brush gear is operated by two rods, one of which is moved by the setting-on shaft and the other by the cam shaft in the knitting frame. The first rod gives adjustable knitting speed, and the second adjustable fashioning speed. Endwise movement of the setting-on rod starts or stops the motor. This rod can be calibrated for knitting speed in courses per minute; it can also be labelled with the names of operations such as 'Clocking' or 'Plating' so that the usual speeds for these operations are quickly obtained. The performance is superior to that of any other kind of drive. Fig. 124 shows the motor and control mechanism mounted at the back of the knitting machine.

This method is preferable to earlier methods with which the speed changes were brought about by switching. Owing to the frequent jerks on the brush gear it is better to make the connexions to the brushes through slip-ring contacts rather than with flexible leads.

Machine tools

Certain machine tool drives, particularly for feed motions, where a constant torque is required over a range of speed, provide an important field for commutator motors. A common requirement is that the work shall be brought rapidly to rest when the operation is complete.

The shunt motor has been extensively used for driving the workhead on grinding machines; an example is shown in Fig. 125. The motor, which must be of small size, is mounted on the workhead, and the regulator fits inside the frame of the grinding machine. A single control lever, operating a switch, serves to start up the motor to any speed for which the regulator handwheel is pre-set, and to bring the motor rapidly to rest by means of a special dynamic braking connexion, which is made automatically when the control lever is moved to the stop position.

BIBLIOGRAPHY

BOOKS

ARNOLD. *Die Wechselstromtechnik*, vol. v. Berlin: Julius Springer, 1912.
DREYFUS. *Kommutatorkaskaden und Phasenschieber.* Berlin: Julius Springer, 1931.
OLLIVER. *The A.C. Commutator Motor.* London: Chapman and Hall, 1927.
RICHTER. *Elektrische Maschinen*, vol. v. Berlin: Julius Springer, 1950.
SCHENKEL. *Die Kommutatormaschinen.* Berlin: de Gruyter, 1924.
TEAGO. *The Commutator Motor.* London: Methuen, 1930.
WALKER. *Specification and Design of Dynamo-electric Machinery.* London: Longmans, 1915.
WALKER. *The Control of the Speed and Power Factor of Induction Motors.* London: Benn, 1924.
ZABRANSKY. *Die Drehzahlregelung von Asynchronmotoren durch Wechselstrom-Kommutator-Hintermaschinen.* Berlin: Heymanns, 1934.

PAPERS

The following list includes a few papers which amplify the subjects dealt with in the book:

ADKINS, B. 'Auxiliary commutating windings in A.C. commutator machines.' *B.T.H. Activities*, vol. XVII, April, 1941, p. 25 (see also *Engineering*, 11 July 1941, p. 26).
ADKINS, B. and GIBBS, W. J. 'Polyphase commutator machines.' *Journal I.E.E.* vol. XCVI, pt. II, 1949, p. 233.
GIBBS, W. J. 'The equations and circle diagrams of the Schrage motor.' *Journal I.E.E.* vol. XCIII, pt. II, 1946, p. 621.
HULL, J. I. 'Theory of speed and power factor control of large induction motors.' *General Electric Review*, vol. XXIII, 1920, p. 630.
SCHRAGE, H. K. '*Mehrfachparallelwicklungen für Drehfeld-Kommutatormaschinen.*' *Swiss (S.E.V.) Bulletin*, vol. XXXIV, 1934, p. 138.
SCHWARZ, B. 'Die neuere Entwicklungen des ständergespeisten Drehstrom-Nebenschluss-Kollektormotors.' *Elektrotechnik und Maschinenbau*, vol. LIII, 1935, p. 85.

INDEX

Fig. 108. Chain-grate stoker driven by Schrage motor

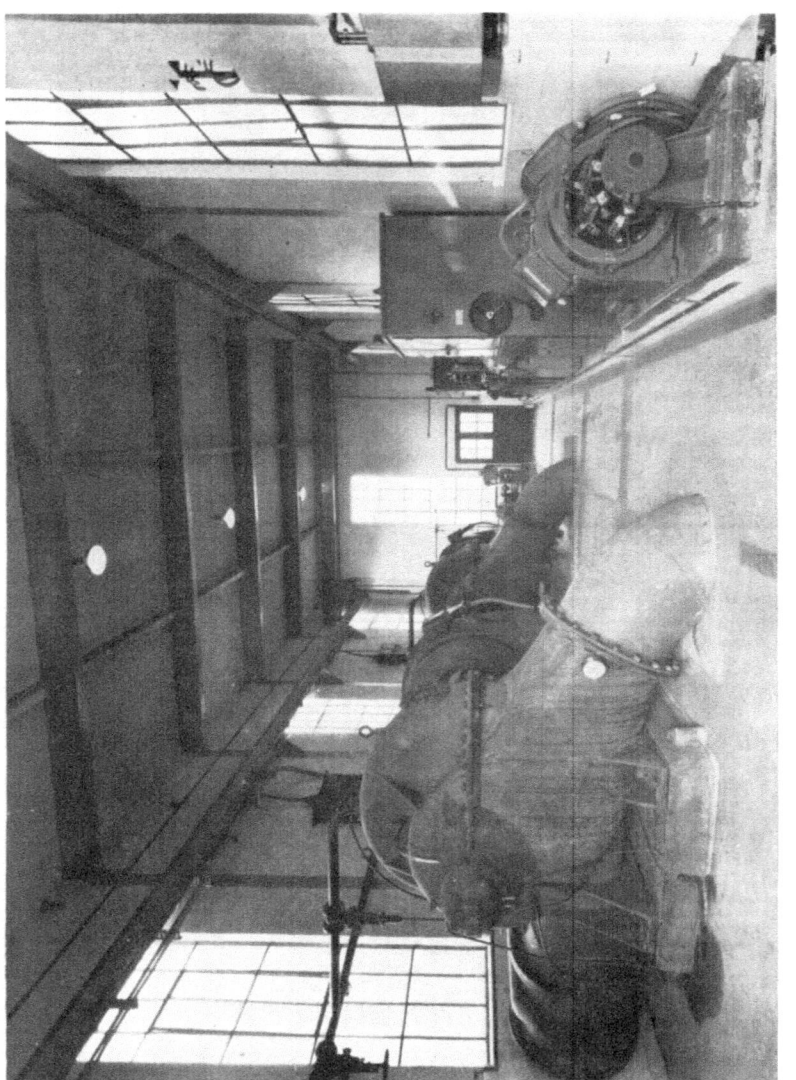

Fig. 109. Circulating pumps driven by single-range Scherbius equipment

Fig. 110. Induction motor and frequency changer for a Scherbius equipment driving a boiler feed pump

Fig. 111. Frequency changer for induction motors driving 'phase changing' pumps

Fig. 112. Large super-calender driven by a Schrage motor

Fig. 113. A.C. sectional drive for paper-making machine

Fig. 114. Scherbius controlled induction motor driving a sheet mill

Fig. 115. Aluminium strip mill driven by a Schrage motor

Fig. 116. Lift installation using Schrage motors

Fig. 117. Overhead crane with selsyn controlled Schrage motor

Fig. 118. Induction motor with Scherbius machine and frequency changer for mine-winder equipment

Fig. 119. Vertical Schrage motor driving
bore-hole pump

Fig. 120. Double-range Scherbius equipment driving a sewage pump

Fig. 121. Shunt motor driving a small printing press

Fig. 122. Large induction motor with shunt phase advancer

Fig. 123. Variable-speed frequency changer set

Fig. 124. Schrage motor driving knitting machine

Fig. 125. Grinding machine with shunt motor and regulator

CAMBRIDGE LIBRARY COLLECTION

Books of enduring scholarly value

History

The books reissued in this series include accounts of historical events and movements by eye-witnesses and contemporaries, as well as landmark studies that assembled significant source materials or developed new historiographical methods. The series includes work in social, political and military history on a wide range of periods and regions, giving modern scholars ready access to influential publications of the past.

The Journals of Walter White

Although he left school at fourteen to work as an upholsterer and cabinet-maker, Walter White (1811–93) would spend forty years working in the library of the Royal Society. White was mostly self-taught, a voracious reader who also learnt German, French, and Latin, and a diligent attender at lectures and other events offering self-improvement. After a brief emigration to the United States, he returned to Britain in 1839, and was offered a post as 'attendant' in the Royal Society's library in 1844; this led to his cataloguing much of the collection, and in 1861 he was appointed Librarian. He became acquainted with many of the Society's members, including Thomas Carlyle, Charles Darwin, and Lord Tennyson. These journals, published posthumously by his brother in 1898, begin with a brief account of his early years before charting his intellectual progress and career, ending in the year he retired, 1884.

Cambridge University Press has long been a pioneer in the reissuing of out-of-print titles from its own backlist, producing digital reprints of books that are still sought after by scholars and students but could not be reprinted economically using traditional technology. The Cambridge Library Collection extends this activity to a wider range of books which are still of importance to researchers and professionals, either for the source material they contain, or as landmarks in the history of their academic discipline.

Drawing from the world-renowned collections in the Cambridge University Library and other partner libraries, and guided by the advice of experts in each subject area, Cambridge University Press is using state-of-the-art scanning machines in its own Printing House to capture the content of each book selected for inclusion. The files are processed to give a consistently clear, crisp image, and the books finished to the high quality standard for which the Press is recognised around the world. The latest print-on-demand technology ensures that the books will remain available indefinitely, and that orders for single or multiple copies can quickly be supplied.

The Cambridge Library Collection brings back to life books of enduring scholarly value (including out-of-copyright works originally issued by other publishers) across a wide range of disciplines in the humanities and social sciences and in science and technology.